BIOSENSORS FOR SINGLE-CELL ANALYSIS

BIOSENSORS FOR SINGLE-CELL ANALYSIS

Edited by

JIAN CHEN
State Key Laboratory of Transducer Technology, Aerospace Information Research Institute, Chinese Academy of Sciences, Beijing, China

YAO LU
Department of Biotechnology, Dalian Institute of Chemical Physics, Chinese Academy of Sciences, Dalian, Liaoning, China

ELSEVIER

ACADEMIC PRESS
An imprint of Elsevier

Academic Press is an imprint of Elsevier
125 London Wall, London EC2Y 5AS, United Kingdom
525 B Street, Suite 1650, San Diego, CA 92101, United States
50 Hampshire Street, 5th Floor, Cambridge, MA 02139, United States
The Boulevard, Langford Lane, Kidlington, Oxford OX5 1GB, United Kingdom

Notices
Knowledge and best practice in this field are constantly changing. As new research and experience broaden our understanding, changes in research methods, professional practices, or medical treatment may become necessary.

Practitioners and researchers must always rely on their own experience and knowledge in evaluating and using any information, methods, compounds, or experiments described herein. In using such information or methods they should be mindful of their own safety and the safety of others, including parties for whom they have a professional responsibility.

To the fullest extent of the law, neither the Publisher nor the authors, contributors, or editors, assume any liability for any injury and/or damage to persons or property as a matter of products liability, negligence or otherwise, or from any use or operation of any methods, products, instructions, or ideas contained in the material herein.

Library of Congress Cataloging-in-Publication Data
A catalog record for this book is available from the Library of Congress

British Library Cataloguing-in-Publication Data
A catalogue record for this book is available from the British Library

ISBN 978-0-323-89841-6

For information on all Academic Press publications
visit our website at https://www.elsevier.com/books-and-journals

Publisher: Andre Gerhard Wolff
Acquisitions Editor: Michelle W. Fisher
Editorial Project Manager: Mica Ortega
Production Project Manager: Omer Mukthar
Cover Designer: Miles Hitchen

Typeset by STRAIVE, India

Contents

Contributors

Ye Ai
Pillar of Engineering Product Development, Singapore University of Technology and Design, Singapore, Singapore

Federica Caselli
Department of Civil Engineering and Computer Science, University of Rome Tor Vergata, Rome, Italy

Deyong Chen
State Key Laboratory of Transducer Technology, Aerospace Information Research Institute, Chinese Academy of Sciences; School of Future Technology; School of Electronic, Electrical and Communication Engineering, University of Chinese Academy of Sciences, Beijing, China

Jian Chen
State Key Laboratory of Transducer Technology, Aerospace Information Research Institute, Chinese Academy of Sciences, Beijing, China

Xingxiu Chen
Department of Chemistry and Biochemistry, Stephenson Life Sciences Research Center, University of Oklahoma, Norman, OK, United States

Jonathan Cottet
Department of Mechanical Engineering, Massachusetts Institute of Technology, Cambridge, MA, United States

Ying Li
Key Laboratory of Magnetic Resonance in Biological Systems, State Key Laboratory of Magnetic Resonance and Atomic and Molecular Physics, National Center for Magnetic Resonance in Wuhan, Wuhan Institute of Physics and Mathematics, Innovation Academy for Precision Measurement Science and Technology, Chinese Academy of Sciences-Wuhan National Laboratory for Optoelectronics, Huazhong University of Science and Technology, Wuhan, Hubei; University of Chinese Academy of Sciences, Beijing, China

Yueying Li
CAS Key Laboratory of Genomic and Precision Medicine, Collaborative Innovation Center of Genetics and Development, Beijing Institute of Genomics, Chinese Academy of Sciences; China National Center for Bioinformation; Savaid Medical School, University of Chinese Academy of Sciences, Beijing, China

Minhui Liang
Pillar of Engineering Product Development, Singapore University of Technology and Design, Singapore, Singapore

Maili Liu
Key Laboratory of Magnetic Resonance in Biological Systems, State Key Laboratory of Magnetic Resonance and Atomic and Molecular Physics, National Center for Magnetic Resonance in Wuhan, Wuhan Institute of Physics and Mathematics, Innovation Academy for Precision Measurement Science and Technology, Chinese Academy of Sciences-Wuhan National Laboratory for Optoelectronics, Huazhong University of Science and Technology, Wuhan, Hubei; University of Chinese Academy of Sciences, Beijing, China

Ziwei Liu
State Key Laboratory of Transducer Technology, Aerospace Information Research Institute, Chinese Academy of Sciences; University of Chinese Academy of Sciences, Beijing, China

Yao Lu
Department of Biotechnology, Dalian Institute of Chemical Physics, Chinese Academy of Sciences, Dalian, Liaoning, China

Zhi-Mei Qi
State Key Laboratory of Transducer Technology, Aerospace Information Research Institute, Chinese Academy of Sciences; University of Chinese Academy of Sciences, Beijing, China

Xuantao Su
School of Microelectronics; Advanced Medical Research Institute, Shandong University, Jinan, Shandong, China

Huiwen Tan
State Key Laboratory of Transducer Technology, Aerospace Information Research Institute, Chinese Academy of Sciences; School of Future Technology, University of Chinese Academy of Sciences, Beijing, China

Junbo Wang
State Key Laboratory of Transducer Technology, Aerospace Information Research Institute, Chinese Academy of Sciences; School of Future Technology; School of Electronic, Electrical and Communication Engineering, University of Chinese Academy of Sciences, Beijing, China

Yunhuang Yang
Key Laboratory of Magnetic Resonance in Biological Systems, State Key Laboratory of Magnetic Resonance and Atomic and Molecular Physics, National Center for Magnetic Resonance in Wuhan, Wuhan Institute of Physics and Mathematics, Innovation Academy for Precision Measurement Science and Technology, Chinese Academy of Sciences-Wuhan National Laboratory for Optoelectronics, Huazhong University of Science and Technology, Wuhan, Hubei; University of Chinese Academy of Sciences, Beijing, China

Zhibo Yang
Department of Chemistry and Biochemistry, Stephenson Life Sciences Research Center, University of Oklahoma, Norman, OK, United States

Ting Zhang
State Key Laboratory of Transducer Technology, Aerospace Information Research Institute, Chinese Academy of Sciences; School of Future Technology, University of Chinese Academy of Sciences, Beijing, China

Shuhong Zhao
Key Laboratory of Magnetic Resonance in Biological Systems, State Key Laboratory of Magnetic Resonance and Atomic and Molecular Physics, National Center for Magnetic Resonance in Wuhan, Wuhan Institute of Physics and Mathematics, Innovation Academy for Precision Measurement Science and Technology, Chinese Academy of Sciences-Wuhan National Laboratory for Optoelectronics; Wuhan National Laboratory for Optoelectronics, Huazhong University of Science and Technology, Wuhan, Hubei, China

Yin Zhao
Key Laboratory of Magnetic Resonance in Biological Systems, State Key Laboratory of Magnetic Resonance and Atomic and Molecular Physics, National Center for Magnetic Resonance in Wuhan, Wuhan Institute of Physics and Mathematics, Innovation Academy for Precision Measurement Science and Technology, Chinese Academy of Sciences-Wuhan National Laboratory for Optoelectronics, Huazhong University of Science and Technology; State Key Laboratory of Hybrid Rice, Institute for Advanced Studies (IAS), Wuhan University, Wuhan, Hubei, China

Jianwei Zhong
Pillar of Engineering Product Development, Singapore University of Technology and Design, Singapore, Singapore

Foreword

Sensitive electronics is a subject that studies and discovers physical, chemical, and biological sensitive characteristics of materials to form sensitive devices and systems to transform human perceptions and knowledge objects into measurable values. It mainly studies new effects, new mechanisms, new methods, and new technologies of information perceptions with key research directions including sensitive mechanisms of various new sensors, characteristics of new sensitive materials, structures and models of new sensitive functions, and new technologies and methods of sensor preparations.

As an interdisciplinary subject, sensitive electronics involves electronics, microelectronics, physics, chemistry, biology, information technologies and other disciplines, functioning as the unique gate of information acquisition. Owing to this key role, sensitive electronics has penetrated into almost all fields such as materials, energy, transportation, aviation, aerospace, public safety, environmental monitoring, medicine, petrochemical, and national defense.

According to different carriers of energy exchanges, main developments of sensitive electronics follow the roadmap of electromechanical energy exchanges, electronic energy exchanges, photoelectric energy exchanges, and quantum energy exchanges. Sensitive devices and systems relying on various modes of energy exchanges coexist for a long time and play unique and important roles in specific application fields. With the characteristics of miniaturization, integration, intellectualization and networking, sensitive electronics will develop from single sensitive devices to multi-sensor integration, sensor network, and microsensors/nanosensors.

As a key branch of sensitive electronics, biosensors are based on biomolecular recognitions with bioactive materials as sensitive elements and life indicators as targets. It is an interdisciplinary frontier of biology, chemistry, physics, materials, information, and bionics. Biosensors have the characteristics of fast, accurate, simple, and high-throughput analysis and thus demonstrate broad application prospects in the fields of biological process control, agriculture and food safety, drug analysis, environmental quality monitoring, biosafety and biosecurity, war agents, life science research, disease diagnosis, and monitoring. The future developments of biosensors will focus on high stability, integration and online measurements, which are expected to bring significant changes to relevant fields.

Different from conventional textbooks covering the whole field of biosensors, in this book, biosensors for single-cell analysis have been explored and analyzed in detail. As an emerging direction of biosensors, biosensors for single-cell analysis have been functioning as a key analytical tool for cell biology and precision medicine. From the perspective of technical developments, recent years have witnessed emergence of exciting biosensors including the "Nature Methods" paper of single-cell mechanical biosensing developed by the Prof. Manalis group from the Massachusetts Institute of Technology in 2019 and the "Nature Methods" paper of single-cell proteomic biosensing developed by the Prof. Nolan group from the Stanford University in 2018. From the perspective of applications, biosensors for single-cell analysis have been used to address major scientific problems in tumor and stem cell biology, and meet key clinical requirements of accurate diagnosis and managements of tumor and blood diseases.

In summary, biosensors for single-cell analysis are one of key research frontiers among scientists and engineers. As there is no previous book to specifically address this topic, it is believed that the book entitled "biosensors for single-cell analysis" can cover in the blanks and provide insights in the relevant areas.

Jiangong Cheng
State Key Laboratory of Transducer Technology, Chinese Academy of Sciences, Shanghai, China

Acknowledgment

The editors acknowledge the tremendous supports of the State Key Laboratory of Transducer Technology of Chinese Academy of Sciences. They are grateful to Michelle Fisher, Mona Zahir, and Omer Mukthar affiliated with Elsevier Inc., who assisted them in the editing and production of the book. Furthermore, the editors deeply appreciate the enduring encouragements and valuable discussions by Profs. Ranran Wang (Shanghai Institute of Ceramics of Chinese Academy of Sciences), Wenhui Wang (Tsinghua University), and Chengjun Huang (Institute of Microelectronics of Chinese Academy of Sciences) in the book draft and publication. They convey their thanks to Lijuan Wang, Jiashuo Chen, and Jianing Chen, family members of Editor J.C.

This book is supported by the National Natural Science Foundation of China (Grant No. 21874133, 31927802, 61922079), the Youth Innovation Promotion Association (Grant No. 2018217, Y201927), and key project (Grant No. QYZDB-SSW-JSC011) of Chinese Academy of Sciences, and the Dalian Institute of Chemical Physics (Grant No. DICP I201908).

CHAPTER 1

Introduction to biosensors for single-cell analysis with historical developments highlighted

Ting Zhang[a,b], Yueying Li[c,d,e], Deyong Chen[a,b,f], Junbo Wang[a,b,f,*], Yao Lu[g,*], and Jian Chen[a,*]

[a]State Key Laboratory of Transducer Technology, Aerospace Information Research Institute, Chinese Academy of Sciences, Beijing, China
[b]School of Future Technology, University of Chinese Academy of Sciences, Beijing, China
[c]CAS Key Laboratory of Genomic and Precision Medicine, Collaborative Innovation Center of Genetics and Development, Beijing Institute of Genomics, Chinese Academy of Sciences, Beijing, China
[d]China National Center for Bioinformation, Beijing, China
[e]Savaid Medical School, University of Chinese Academy of Sciences, Beijing, China
[f]School of Electronic, Electrical and Communication Engineering, University of Chinese Academy of Sciences, Beijing, China
[g]Department of Biotechnology, Dalian Institute of Chemical Physics, Chinese Academy of Sciences, Dalian, Liaoning, China

1 Introduction to biosensors for single-cell analysis

A biosensor is a device used to sense a biological variable, which includes, but is not limited to a type/concentration of a specific analyte (a biochemical compound, a sequence of nucleic acid, a specific protein, virus particle, and a bacterium). To sense these variables, a biosensor is usually composed of a bioreceptor and a transducer, where a bioreceptor is made of antibodies, enzymes, nucleic acids, viruses, bacteria, cells or tissues, functioning to specifically bind with the target analyte. Then a transducer converts the captured analyte into a universal signal (e.g., a voltage), which mainly relies on electrochemical (by measuring voltage or current), optical (by measuring light intensity), thermal (by measuring temperature) or piezoelectric (by measuring mass) mechanisms.

As a key type of analytes, isolated biological cells are considered as a key intermediate between complex studies in living animals and simple biomolecular studies in well-defined environments. Thus biosensors for cell characterization have emerged as indispensable tools for fundamental cell

*Co-Corresponding authors.

Biosensors for Single-Cell Analysis
https://doi.org/10.1016/B978-0-323-89841-6.00007-4

biology aimed to unravel the yet unknown secrets of life, for drug developments and for screening xenobiotics for their impacts on living organisms. However, the majorities of cells are analyzed in population without effectively addressing the issue of cellular heterogeneity, which plays more and more important roles in cell biology and precision medicine. To meet these requirements, biosensors for single-cell analysis have been under research focus to address major scientific problems in tumor and stem cell biology, and meet key clinical requirements including accurate diagnosis and managements of tumor and blood diseases (Table 1).

2 Historical developments of biosensors for single-cell analysis

Biosensors of single-cell analysis can be traced back to 1937 when Curtis and Cole placed a *Nitella* cell in a groove embedded by two electrodes with collected single-cell electrical properties reported.[1] Early developments of biosensors for single-cell analysis were closely associated with glass pipettes for cell manipulation and position. In 1964 a prototype of micropipette aspiration was demonstrated where a red blood cell was sucked into a glass micropipette with corresponding deformations recorded and translated into mechanical properties.[3] In 1965 an application of capillary electrophoresis for single-cell analysis was demonstrated where hemoglobin mixtures from single erythrocytes were separated into individual constituents based on electrophoretic forces.[4] Furthermore, in 1968, a prototype of patch clamping was demonstrated where isolated erythrocytes in suspension were penetrated by a glass micropipette filled with ionic solutions to probe membrane potentials.[6]

Further technical improvements of cell manipulations pushed forward the emerging approaches of biosensors for single-cell analysis. More specifically, in 1978, mass spectroscopy was used for single-cell analysis where a laser beam with a few micrometers in diameter probed the chemistry of thin tissue slices and various metabolites were detected in different regions of single cells.[8] In 1981, a rotating electric field was exerted to rotate a suspended single cell as a result of Maxwell-Wanger polarization and the corresponding rotating speeds were translated into electrical properties.[9] In 1987, microcolumn liquid chromatography and electrochemical analysis were used for quantification of single-neuron metabolic properties.[10] In 1995, optical tweezers were used for single-cell characterization, where multiple laser lights were shined on a red blood cell and the generated light pressure caused cellular deformation with single-cell mechanical properties measured.[13]

Table 1 Key developments in the history of biosensors for single-cell analysis.

Year	Parameter of biosensor	Type of biosensor	References
1937	Electrical property	Single cell between two electrodes	1
1956	Electrical property	Electrical flow cytometry	2
1964	Mechanical property	Micropipette aspiration	3
1965	Proteomic property	Capillary electrophoresis	4
1965	Optical property	Optical flow cytometry	5
1968	Electrical property	Patch clamping	6
1972	Proteomic property	Protein flow cytometry	7
1978	Metabolic property	Mass spectroscopy	8
1981	Electrical property	Electrorotation	9
1987	Metabolic property	Microcolumn liquid chromatography	10
1992	Electrical property	On-chip electrorotation	11
1992	Mechanical property	On-chip deformability flow cytometry	12
1995	Mechanical property	Optical tweezer	13
1997	Metabolic property	On-chip capillary electrophoresis	14
1997	Electrical property	On-chip impedance flow cytometry	15
1999	Proteomic property	On-chip fluorescent flow cytometry	16
2000	Electrical property	On-chip patch clamping	17
2004	Optical property	Image flow cytometry	18
2009	Proteomic and metabolic properties	Mass flow cytometry	19
2011	Mechanical property	On-chip micropipette aspiration	20

Within the past 30 years, miniaturization of conventional biosensors for single-cell characterization is a leading trend due to development of lab-on-a-chip technologies. In 1992, 1997, 2000, and 2011, on-chip electrorotation,[11] on-chip capillary electrophoresis,[14] on-chip patch clamping,[17] and on-chip micropipette aspiration[20] were reported respectively, as key milestones in the developments of biosensors enabling single-cell characterization.

Besides accurate manipulation and positioning of single cells, a second approach of biosensors for single-cell analysis is to flush single cells one-by-one through a detection domain with corresponding properties measured, which is termed as flow cytometry. Although it is difficult to estimate a single starting point of flow cytometry, the main originality residing in observation of aligned cells one behind another into a fluid sheath can be traced back to 1934.[21] The first flow cytometer capable of measuring the electrical properties of single particles was developed by Coulter (1956) where the device measured the resistance as single cells passed through a small orifice.[2] In 1965, single cells were forced to pass across a laser beam with scattered light quantified as a prototype of optical flow cytometry.[5] Furthermore, in 1972, Bonner et al. developed a flow cytometry to analyze FITC- and Rhodamine-coupled antibody fluorescence as a prototype demonstration of proteomic flow cytometry.[7]

Recently, there is a clear trend to miniaturize conventional flow cytometry, where flow channels based on microfabrication were used to replace conventional flow channels enabling accurate geometrical positioning of cells under measurements. Key developments in microfluidic flow cytometry mainly include but are not limited to on-chip deformability flow cytometry in 1992,[12] on-chip impedance flow cytometry in 1997,[15] and on-chip fluorescent flow cytometry.[16] Meanwhile, mass and image flow cytometry were also developed as enabling instruments for single-cell analysis that allow characterization of multiple simultaneous cellular and subcellular parameters. As a prototype demonstration of image flow cytometry, in 2004 high-resolution light and fluorescent microscopy were combined to capture an image for each event passing through the detector, enabling the characterization of single-cell optical properties.[18] As to mass flow cytometry, in 2009, rare earth metal isotopes were used to label cells that were then read by a mass spectrometer, allowing for multiple characterization of single-cell proteomic and metabolic properties.[19]

3 Book structure

In this book, biosensors for single-cell analysis are classified into biosensors for the characterization of single-cell biochemical analysis including proteomic and metabolic properties and single-cell biophysical analysis including electrical, mechanical, and optical properties. Then, follows a short introduction to the book. The first chapter gives an introduction to biosensors for single-cell analysis with historical developments. The two subsequent chapters deal with biosensors enabling biochemical (proteomic and metabolic) characterization of single cells; Chapters 4–6 describe the developments of biosensors enabling the measurements of single cells from biophysical perspectives including electrical, mechanical, and optical properties. The last chapter functions as a summary and suggests future directions of biosensors for single-cell analysis from the perspective of instrumentation.

Acknowledgment

The authors acknowledge financial supports from the National Natural Science Foundation of China (Grant No. 61825107, 61922079, 62001042), Key Project (QYZDB-SSW-JSC011), Instrument Development, Youth Innovation Promotion Association, and Interdisciplinary Innovation Team of Chinese Academy of Sciences.

References

1. Curtis, H. J.; Cole, K. S. Tranverse Electric Impedance of Nitella. *J. Gen. Physiol.* **1937**, *21* (2), 189–201.
2. Coulter, W. H. High Speed Automatic Blood Cell Counter and Cell Size Analyzer. In Proceedings of the National Electronics Conference, Vol. 12; 1956; p. 1034.
3. Rand, R. P.; Burton, A. C. Mechanical Properties of the Red Cell Membrane. I. Membrane Stiffness and Intracellular Pressure. *Biophys. J.* **1964**, *4* (2), 115–135.
4. Matioli, G. T.; Niewisch, H. B. Electrophoresis of Hemoglobin in Single Erythrocytes. *Science* **1965**, *150* (3705), 1824–1826.
5. Kamentsky, L. A.; Melamed, M. R.; Derman, H. Spectrophotometer: New Instrument for Ultrarapid Cell Analysis. *Science* **1965**, *150* (3696), 630–631.
6. Lassen, U. V.; Sten-Knudsen, O. Direct Measurements of Membrane Potential and Membrane Resistance of Human Red Cells. *J. Physiol.* **1968**, *195* (3), 681–696.
7. Bonner, W. A.; Hulett, H. R.; Sweet, R. G.; Herzenberg, L. A. Fluorescence Activated Cell Sorting. *Rev. Sci. Instrum.* **1972**, *43* (3), 404–409.
8. Kaufmann, R.; Hillenkamp, F.; Nitsche, R.; Schürmann, M.; Wechsung, R. The Laser Microprobe Mass Analyser (LAMMA): Biomedical Applications. *Microsc. Acta Suppl.* **1978**, (2), 297–306.
9. Zimmermann, U.; Vienken, J.; Pilwat, G. Rotation of Cells in an Alternating Electric Field: The Occurrence of a Resonance Frequency. *Z. Naturforsch. C Biosci.* **1981**, *36* (1–2), 173–177.
10. Kennedy, R. T.; St. Claire, R. L.; White, J. G.; Jorgenson, J. W. Chemical Analysis of Single Neurons by Open Tubular Liquid Chromatography. *Microchim. Acta* **1987**, *92* (1), 37–45.
11. Huang, Y.; Hölzel, R.; Pethig, R.; Wang, X. B. Differences in the AC Electrodynamics of Viable and Non-viable Yeast Cells Determined Through Combined Dielectrophoresis and Electrorotation Studies. *Phys. Med. Biol.* **1992**, *37* (7), 1499.
12. Kikuchi, Y.; Sato, K.; Ohki, H.; Kaneko, T. Optically Accessible Microchannels Formed in a Single-Crystal Silicon Substrate for Studies of Blood Rheology. *Microvasc. Res.* **1992**, *44* (2), 226–240.

13. Bronkhorst, P. J.; Streekstra, G. J.; Grimbergen, J.; Nijhof, E. J.; Sixma, J. J.; Brakenhoff, G. J. A New Method to Study Shape Recovery of Red Blood Cells Using Multiple Optical Trapping. *Biophys. J.* **1995**, *69* (5), 1666–1673.
14. Li, P. C.; Harrison, D. J. Transport, Manipulation, and Reaction of Biological Cells On-Chip Using Electrokinetic Effects. *Anal. Chem.* **1997**, *69* (8), 1564–1568.
15. Larsen, U. D.; Blankenstein, G.; Branebjerg, J. Microchip Coulter Particle Counter. In Proceedings of International Solid State Sensors and Actuators Conference (Transducers '97), Vol. 2; **1997**; pp. 1319–1322.
16. Fu, A. Y.; Spence, C.; Scherer, A.; Arnold, F. H.; Quake, S. R. A Microfabricated Fluorescence-Activated Cell Sorter. *Nat. Biotechnol.* **1999**, *17* (11), 1109–1111.
17. Fertig, N.; Tilke, A.; Blick, R. H.; Kotthaus, J. P.; Behrends, J. C.; Bruggencate, G. T. Stable Integration of Isolated Cell Membrane Patches in a Nanomachined Aperture. *Appl. Phys. Lett.* **2000**, 77 (8), 1218–1220.
18. George, T. C.; et al. Distinguishing Modes of Cell Death Using the ImageStream Multispectral Imaging Flow Cytometer. *Cytometry A* **2004**, *59* (2), 237–245.
19. Bandura, D. R.; et al. Mass Cytometry: Technique for Real Time Single Cell Multitarget Immunoassay Based on Inductively Coupled Plasma Time-of-Flight Mass Spectrometry. *Anal. Chem.* **2009**, *81* (16), 6813–6822.
20. Chen, J.; et al. A Microfluidic Device for Simultaneous Electrical and Mechanical Measurements on Single Cells. *Biomicrofluidics* **2011**, *5* (1), 014113.
21. Moldavan, A. Photo-Electric Technique for the Counting of Microscopical Cells. *Science* **1934**, *80* (2069), 188–189.

CHAPTER 2

Biosensors for single-cell proteomic characterization

Yin Zhao[a,b], Shuhong Zhao[a,c], Ying Li[a,d,*], Yunhuang Yang[a,d], and Maili Liu[a,d]

[a]Key Laboratory of Magnetic Resonance in Biological Systems, State Key Laboratory of Magnetic Resonance and Atomic and Molecular Physics, National Center for Magnetic Resonance in Wuhan, Wuhan Institute of Physics and Mathematics, Innovation Academy for Precision Measurement Science and Technology, Chinese Academy of Sciences-Wuhan National Laboratory for Optoelectronics, Huazhong University of Science and Technology, Wuhan, Hubei, China
[b]State Key Laboratory of Hybrid Rice, Institute for Advanced Studies (IAS), Wuhan University, Wuhan, Hubei, China
[c]Wuhan National Laboratory for Optoelectronics, Huazhong University of Science and Technology, Wuhan, Hubei, China
[d]University of Chinese Academy of Sciences, Beijing, China

1 Introduction

Cells are the building blocks of multicellular living organism. All human cells are originated from a single zygote. However, numerous studies have revealed that even the isogenic cells demonstrate significant differences in transcriptome, metabolome, and proteome.[1–3] Traditional bulk cell measurements could not reveal the heterogeneity of individual cells and thus may mask the important function of some spare cells, which might be the key for understanding the mechanism of physiological processes.[4,5] Recent advances in single-cell analysis (SCA) have uncovered previously unappreciated heterogeneities in cell populations that seem to be uniform, enabling the classification of cells into distinct subtypes based on their defining characteristics.[6,7] Single-cell based studies has also been revolutionizing the clinical medicine.[8–10] For example, cancer immunotherapy and hematologic oncology are two examples where single cell-based analysis provides valuable insights toward effective personalized therapies.[11]

"Proteome" is considered as the overall proteins or peptides of a cell that reflect their localizations, interactions, and posttranslational modifications at a specific time.[12] The term of "proteomics" was first proposed in 1996 to denote the "PROTein complement of a genOME."[13] Proteomics represents the functional information of genes, and it provides vital clues for

[*]Corresponding author.

https://doi.org/10.1016/B978-0-323-89841-6.00004-9

biochemical processes and signaling pathways.[14, 15] Specifically, single cell proteomics (SCP) could reveal the proteome differences between individual cells, which will further facilitate our understanding on many physiological processes such as stem cell differentiation, immune responses, cancer metastasis, tumor cell responses to drugs, and others.[16–18]

In comparison with genomics, SCP is more challenging to elucidate because of the following reasons: (1) Complexity. It is estimated per protein-coding gene (~20,000 in human cells) produces more than 100 protein isoforms.[12, 19] What makes the situation even more complex is the posttranslational modifications (>300 known modifications), including the most common modifications of glycosylation, phosphorylation, methylation, and acetylation.[20] Additionally, there are complex interactions between these proteins. (2) Dynamics. The proteome fluctuates from time to time and varies when cells are exposed to external stimuli.[21] (3) Proteins cannot be amplified. In a typical human cell, the proteome can cover a three-order range in concentration from a few hundred copies to >100,000 copies.[3, 4] However, there are no strategies to directly amplify proteins, rendering the quantification of low abundance proteins significantly difficult.

Although significant difficulties are present, SCP attracts great interests from the research community. To measure single cell proteome, many platforms have been proposed based on various strategies.[5, 16, 22] For example, 96-well microplates or Petri dishes, combined with a microscopy, are common tools traditionally used for single cell proteome analysis based on fluorescence imaging.[23] This type of measurement has a high spatial resolution to show the location of the target proteins, but it suffers from low throughput and requires laborious work for cell locating and focusing. Flow cytometry is a high-throughput platform (>1000 cells per second) for SCP analysis based on fluorescence-activated cell sorting (FACS).[24] FACS is able to analyze ~10 proteins at a time, by using fluorophore-labeled antibodies.[25] FACS has produced valuable information for the advancement of SCP. However, the protein targets are difficult to further increase due to the spectra overlap of the fluorophores.[26, 27] Mass cytometry, a new instrument leveraging the features of flow cytometry and mass spectroscopy (MS), has been developed recently for SCA.[26] In contrast to flow cytometry using fluorophores as reporters, mass spectrometry utilizes antibodies coupled to unique heavy-metal isotopes to quantify the target proteins with little signal overlap. Thus mass cytometry is capable of measuring over 40 cellular parameters at single-cell resolution, greatly augmenting the ability of cytometry to investigate complex processes in cells.[28]

Indeed, MS has routinely been used for SCP analysis in the recent decade.[3, 19] MS techniques and systems have been significantly improved with respect to throughput, sensitivity, reproducibility, and proteome coverage.[26, 29]

The technology has been advanced to a state where analyzing low-copy number proteins (~1000 copies) in single cells becomes possible.[3] In contrast with antibody-based methods, the most significant advantage of MS-based methods is the ability of quantifying hundreds of proteins in a single cell, which is extremely meaningful for understanding the interactions and functions of proteins at single-cell resolution. However, a significant challenge for MS-based SCP characterization is sample preparation, which includes cell separation, cell lysis, protein digestion and delivery to MS.[9,29] Owing to the minimal size of a cell (~10 μm for human cell) and the limited quantity of proteins (~500 pg protein in total),[20] it is crucial to choose an appropriate sampling approach for MS to achieve a high throughput and high sensitivity detection for SCP analysis. With the advantages of cell-comparable micro-features, precise fluid controlling ability, parallel sample operation and ease of automation, microfluidics-based platforms have been proved to be effective tools for SCA.[30–33] Microchips are also widely used for single cell sampling and integrated in MS-based platforms for SCP charaterization.[34,35]

Biosensor is recognized as an analytical device used for the detection or characterization of biological elements, including intracellular proteins, cell receptors, enzymes, antibodies, nucleic acids, etc.[36] In this chapter, we mainly focus on the implement of microfluidics-based biosensors for SCP characterization. Approaches such as well plate, flow cytometry, mass cytometry will not be reviewed in the following sections. However, micro-device assisted MS analysis on SCP will be covered. First, we highlight three types of microfluidics strategies for quantifying SCP (Fig. 1): (1) antibody-based

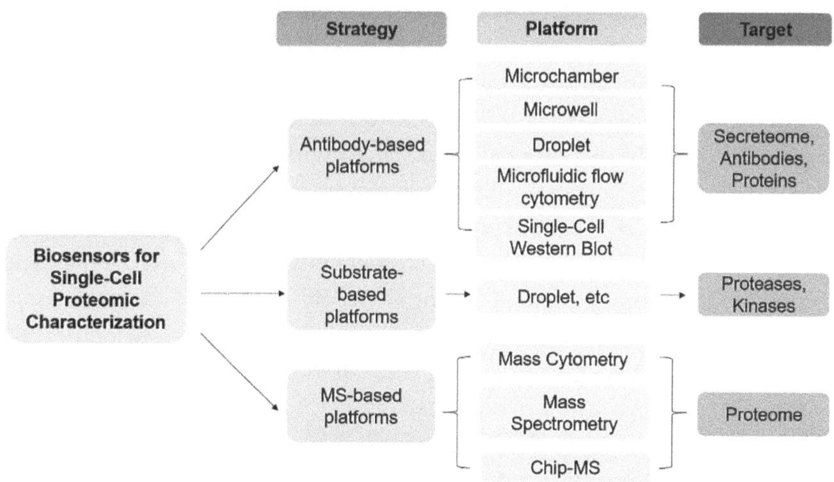

Fig. 1 Overview of the biosensing strategies and platforms for single-cell proteomic characterization.

methods, (2) substrate-based methods, and (3) Chip-MS-based methods. Additionally, we also summarize some commercialized microfluidic systems that are developed for SCP characterization. At last, we make a brief conclusion and propose several future opportunities that could advance the biosensor-based SCP analysis.

2 Antibody-based SCP characterization by microfluidic biosensor

Immunoassay can help to detect diverse proteins both in vitro and in vivo by using the specific antibody to recognize the antigen. Combining microfluidic biosensors with fluorescent probes labeled antibodies could achieve high sensitivity, high-throughput, multiplexing, and single-cell proteome profiling. In this section, we divide the antibody-based microfluidic biosensing platforms into five categories, which are microchamber, microwell, droplet, microfluidic flow cytometry, and single-cell western blot (Fig. 1).

2.1 Microchamber-based biosensor

Microchamber-based microfluidics is a useful approach for SCP analysis.[37,38] This type of platform is often coupled with pressure-controlled membrane valves to control the flow and cells.[39] Heath's group developed a microchamber-based platform called Single Cell Barcode Chip (SCBC) to quantitatively detect the intracellular signaling proteins of single cancer cells (Fig. 2A).[38] SCBC is comprised of a two-layer microfluidic network, with the top valve layer to isolate the bottom microchannel layer into 120 microchambers for cell compartmentalization, cell lysis, and protein assays. A single or limited number of cells were isolated into the microchamber that were precoated with an antibody array for capturing and detecting a group of proteins. A barcode of 11 antibodies were selected targeting intracellular proteins directly or potentially associated with PI3K signaling. Cells of three isogenic glioblastoma multiforme cell lines of U87, U87 EGFRvIII, and U87 EGFRvIII PTEN were studied in this study. The measured protein abundances reveal single-cell heterogeneity, and different types and strengths of protein-protein interactions. SCBC is able to quantify more than 10 proteins for more than 100 cells in a single device, which could provide a comprehensive picture for understanding cellular signaling networks. The same group also described a chemical approach in SCBC to perform integrated metabolic and proteomic assays from single cells.[40] The results displayed drug-induced changes in the metabolite-phosphoprotein correlation network in single cells.

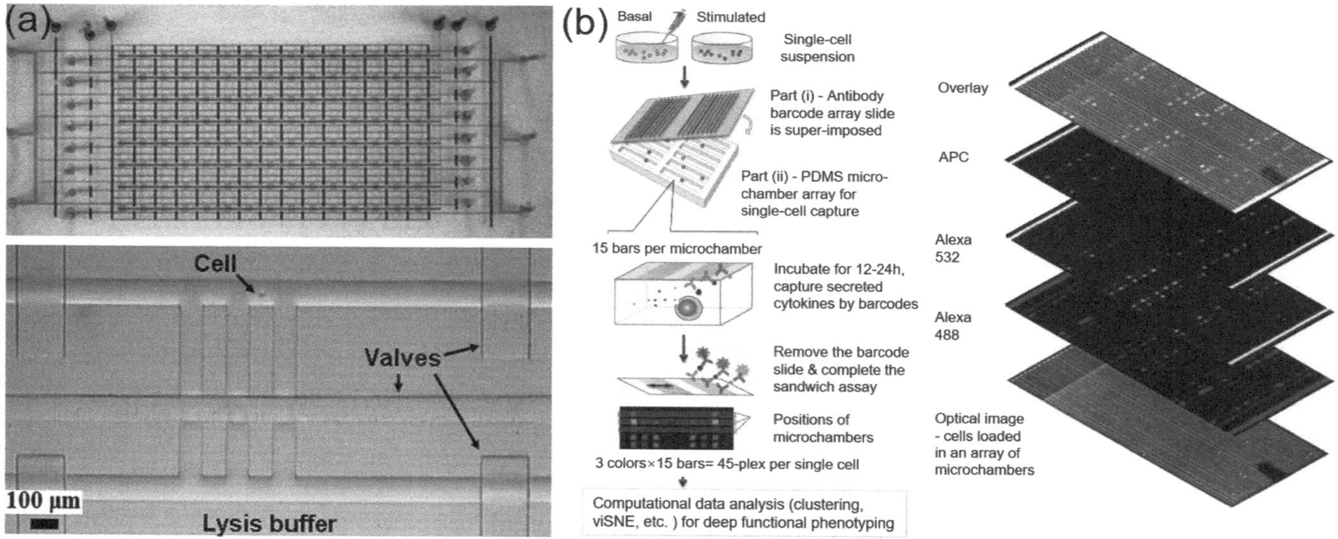

Fig. 2 Microchamber-based platforms for SCP. (A) The single-cell barcode chip (SCBC). (B) Scheme and work flow of SCBC-based multiplexing protein secretion assay. *((A) Reproduced with permission from Shi, Q.; Qin, L.; Wei, W.; Geng, F.; Fan, R.; Shin, Y. S.; Guo, D.; Hood, L.; Mischel, P. S.; Heath, J. R. Single-Cell Proteomic Chip for Profiling Intracellular Signaling Pathways in Single Tumor Cells. Proc. Natl. Acad. Sci. U. S. A.* ***2012,*** *109 (2), 419–424. (B) Reproduced with permission from Lu, Y.; Xue, Q.; Eisele, M. R.; Sulistijo, E. S.; Brower, K.; Han, L.; Amir el, A. D.; Pe'er, D.; Miller-Jensen, K.; Fan, R. Highly Multiplexed Profiling of Single-Cell Effector Functions Reveals Deep Functional Heterogeneity in Response to Pathogenic Ligands. Proc. Natl. Acad. Sci. U. S. A.* ***2015,*** *112 (7), E607–E615.)*

To increase the throughput for analyzing single cell proteome, Lu and Fan et al. reported a microfluidic platform that integrated a large array of subnanoliter microchambers and high-density antibody barcode microarrays (Fig. 2B).[41] A panel of 14 secretory proteins (including cytokines, chemokines, and growth factors involved in immunological or pathophysiological processes) were measured for more than 1000 single cells in parallel. Both cell lines and primary malignant brain tumor cells were tested and distinct heterogeneity among the single cell secretomic signatures were observed. A similar platform developed by the same group was used to interrogate the tumor-macrophage paracrine signaling at single cell level.[42] Additionally, Fan's group improved their platform by combining spectral (color) and spatial (spots) multiplexing to dramatically increase the number of proteins that can be detected for single cells.[43] A total of 42 immune effector proteins secreted from single cells were profiled to analyze the differentiated macrophages stimulated with lipopolysaccharide (LPS). Unique protein profiles were observed for the same single cells before and after LPS stimulation. Significant functional heterogeneity and varying levels of pathogenic activation were also observed among the cells. Extracellular vesicles (EVs) act as important mediators for intercellular communication.[44] Ji and Lu et al. developed a spatially patterned antibody barcode chip and realized multiplexed profiling of single-cell EV secretion for more than 1000 single cells.[45] Human oral squamous cell carcinoma (OSCC) cell lines and primary OSCC cells were analyzed, and it was found that the decrement of certain EV phenotypes (e.g., $CD63^+$ EV) was associated with the invasive feature of these cells. This technology could become a useful tool for the evaluation of EV secretion heterogeneity at single-cell level.

2.2 Microwell-based biosensor

Microwell-based devices are often designed as roofless and with an array of miniatured containers.[46,47] Cells with an appropriate density are loaded into the microwells by gravity.[48] In 2006, Love et al. described a microengraving method that used a dense array of microwells to isolate individual cells and measure molecules secreted by each cell (Fig. 3A).[49] To detect the secreted antibodies by hybridomas, secondary antibodies were covalently immobilized on a glass slide that was applied to seal each well. After 2h incubation, the slide was interrogated with a mixture of fluorescently labeled antigens. This microengraving method is simple, fast and can screen a variety of antibodies for large numbers of individual cells, which represents significant advantages over traditional ELISA. The microengraving method

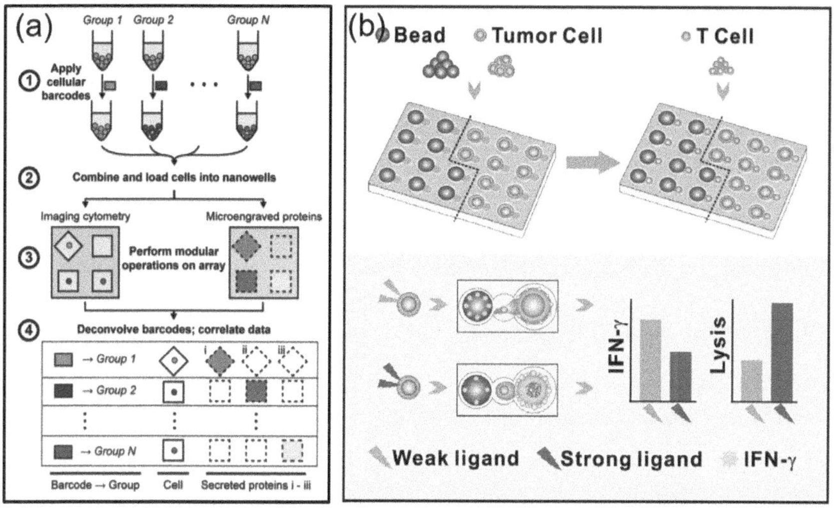

Fig. 3 Microwell-based platform for SCP. (A) Scheme to show the integration of microengraving and fluorescent cellular barcodes for single cell secretome measurement. (B) A hierarchical loading microwell chip enabling simultaneous detection of T-cell cytotoxicity and cytokine secretion. *((A) Reproduced with permission from Yamanaka, Y. J.; Szeto, G. L.; Gierahn, T. M.; Forcier, T. L.; Benedict, K. F.; Brefo, M. S.; Lauffenburger, D. A.; Irvine, D. J.; Love, J. C. Cellular Barcodes for Efficiently Profiling Single-Cell Secretory Responses by Microengraving. Anal. Chem.* ***2012,*** *84 (24), 10531–10536. (B) Reproduced with permission from Zhou, Y.; Shao, N.; Bessa de Castro, R.; Zhang, P.; Ma, Y.; Liu, X.; Huang, F.; Wang, R. F.; Qin, L. Evaluation of Single-Cell Cytokine Secretion and Cell-Cell Interactions with a Hierarchical Loading Microwell Chip. Cell Rep.* ***2020,*** *31 (4), 107574.)*

was also used to quantify the secretion rates of four cytokines released from individual primary T cells.[50] To increase the number of unique samples that can be analyzed simultaneously by microengraving, the same group introduced fluorescent cellular barcodes for quantifying the secretory responses of thousands of single cells in parallel. Based on this strategy, the secretory behavior of mechanistically treated peripheral blood mononuclear cells (PBMCs) and different lineages of $CD4^+$ T-helper cells were profiled.[51]

Multiplexed detection of proteins secreted by single cells is meaningful but challenging. A microwell-based platform was developed with single-stranded DNA encoded microbead arrays and multicolor successive imaging for the characterization of single-cell secreted proteins.[52] Ten immune effector proteins from single macrophages stimulated with LPS were measured and significant heterogeneity was observed. Meanwhile, characterization of T-cell functions (e.g., cytokine secretion and cytolysis of target cells) at the single-cell level is of great importance in adoptive cell therapy.[53]

To achieve a high trapping efficiency for T cell, target cell, and microbeads, a hierarchical loading microwell chip (HL-Chip) that aligns multiple cells was developed by Zhou et al.[53] HL-Chip enables the loading of different objects with different sizes (Fig. 3B). By using this platform, single T-cell functions were investigated in three applications, including longitudinal profiling of IFN-g secretion of single T cells, large-scale evaluation of T-cell cytolytic activity, and T-cell-tumor cell interactions. HL-Chip could be a useful tool for SCP due to its simplicity, scalability, versatility, and precision.

2.3 Droplet-based biosensor

Droplet system is a mainstream of microfluidics-based platforms.[54, 55] It allows the encapsulation of single cells, reagents, and the proteins secreted by the cells or cell lysates in a water-in-oil emulsion droplet with a picoliter or nanoliter volume.[56] The significant advantage of droplet systems is their high throughput, which can achieve a speed of up to thousands of cells per second.[39, 57] In addition, the isolated small compartment allows stable confining of target proteins for a long time, which is suitable for the measurement of proteases or antibodies secreted by cells.[4, 11]

Mazutis et al. demonstrated a droplet-based biosensor for the detection of antibodies secreted from single mouse hybridoma cells.[58] Single hybridoma cells, a fluorescent probe and single beads coated with anti-mouse IgG antibodies were encapsulated in 50-pL droplets. The droplets produced bright fluorescence when enough secreted antibodies bound to the beads. This system can screen ~1 million cells in ~4 h. A similar droplet system named DropMap was developed to characterize antibody-secreting cells from tetanus toxoid immunized mice.[59] The secretion rate and affinity of IgG were simultaneously analyzed based on the fluorescence relocation strategy for 0.5 million cells enriched from spleen and bone marrow. These fluorescence-based approaches were sensitive, but can only detect limited target proteins. To expand the power of droplet approach for measuring multiple targets, a method termed Abseq was proposed to profile multiple proteins in single cells.[60] Abseq uses DNA sequence tags labeled antibodies to recognize epitopes of interest and the readout was conducted by microfluidic barcoding and DNA sequencing. This DNA-tagged antibody approach enables sensitive detection of low-abundance proteins, and shows great potential for multiplexing. Droplet-based biosensor can also be used to perform functional screening of T-cell receptor (TCR) for T cells on recognition of target tumor cells (Fig. 4A).[61] T cell and the target cell were compartmented into single droplets, which were then distributed into an

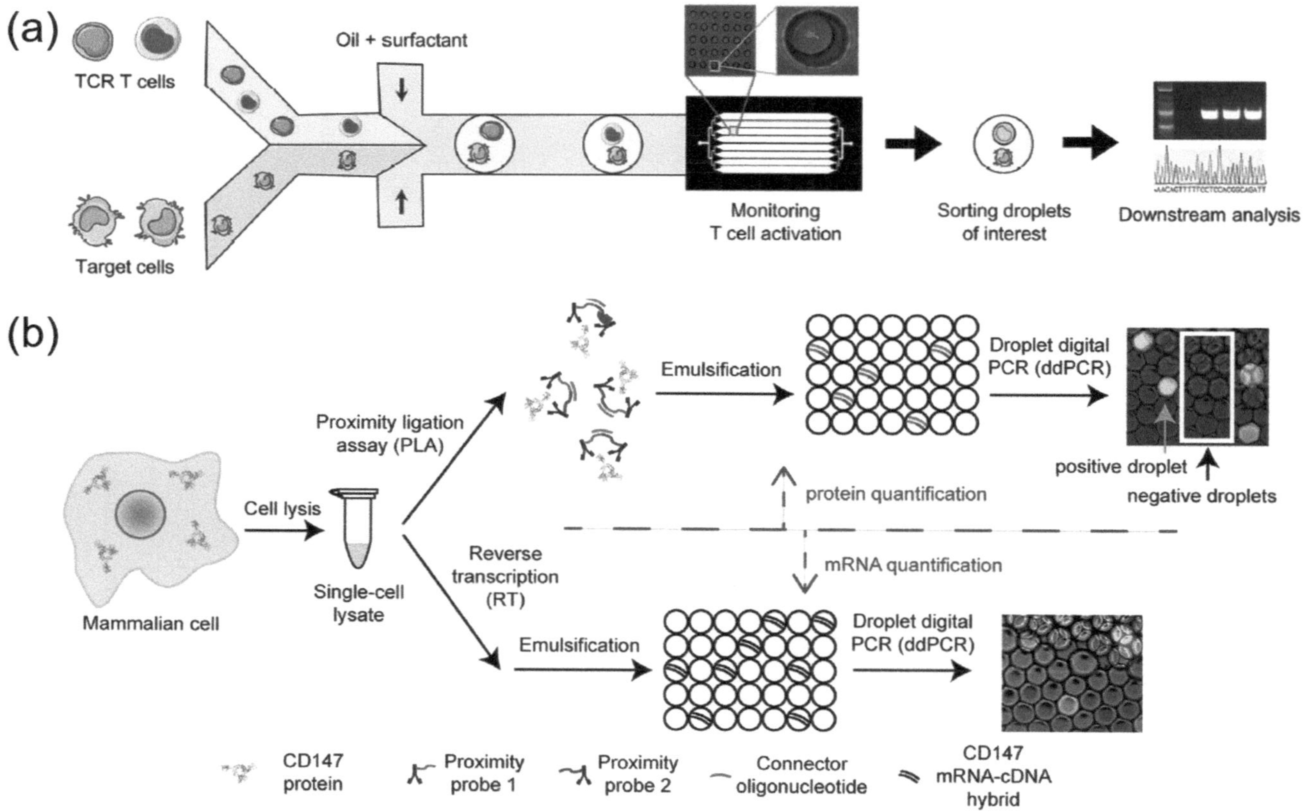

Fig. 4 Droplet-based systems for SCP. (A) Scheme of the droplet-based approach for functional screening of TCR T cells. (B) Illustration of the digital PLA protocol for single cell protein and mRNA quantification. *((A) Reproduced with permission from Segaliny, A. I.; Li, G.; Kong, L.; Ren, C.; Chen, X.; Wang, J. K.; Baltimore, D.; Wu, G.; Zhao, W. Functional TCR T Cell Screening Using Single-Cell Droplet Microfluidics. Lab Chip* ***2018,*** *18 (24), 3733–3749. (B) Reproduced with permission from Albayrak, C.; Jordi, C. A.; Zechner, C.; Lin, J.; Bichsel, C. A.; Khammash, M.; Tay, S. Digital Quantification of Proteins and mRNA in Single Mammalian Cells. Mol. Cell* ***2016,*** *61 (6), 914–924.)*

inverted floating array containing tens of thousands of trapping wells. T-cell activation kinetics was monitored via eGFP expression and the cells of interest could be retrieved for downstream sequence analysis.

Sensitive and absolute quantification of both proteins and mRNA in single cells is critical for understanding and modeling biological systems.[62] Recently, Albayrak et al. reported a workflow that combines proximity ligation assay (PLA) and digital PCR in a droplet platform (Fig. 4B).[63] PLA converts the targeted protein to dsDNA via a pair of oligonucleotide-bound antibodies.[64] The authors used droplet digital PCR (ddPCR) to achieve digital PLA, which has femtomolar sensitivity and enables the quantification of proteins with concentration covering a three-order dynamic range. Based on this strategy, the correlation between CD147 protein and its mRNA were examined for hundreds of single cells. Similarly, Stoeckius et al. developed a droplet platform termed cellular indexing of transcriptomes and epitopes by sequencing (CITE-seq). CITE-seq uses oligonucleotide-labeled antibodies to characterize cellular protein and transcriptome in single cells.[1] A panel of 13 well-characterized monoclonal antibodies were prepared to recognize cell-surface proteins for immune-cell classification. The results demonstrated that multimodal data from CITE-seq enhanced the characterization of cellular phenotypes compared with scRNA-seq alone. This platform could also be very useful for the studies of posttranscriptional gene regulation at the single-cell level.

2.4 Microfluidic flow cytometry

Flow cytometry and mass cytometry have been widely used in single-cell analysis for measuring both surface and intracellular proteins.[25,26] However, these two systems are still difficult to obtain absolute copy numbers of intracellular proteins owing to the lack of calibration strategies.[65,66] To quantify exact protein copy numbers in single cells, a platform termed microfluidic flow cytometry was developed.[66] Single cells stained with fluorophore-labeled antibodies are forced to deform through a constriction channel with a cross-sectional area smaller than a cell, which renders the fluorescence detection domain to be fully occupied. Then, the cellular fluorescent intensities were recorded by a photomultiplier tube (PMT) as the raw data. Next, a series of FITC-labeled antibody solutions with different concentrations was flushed through the constriction area and the fluorescent intensities were recorded to produce the calibration curve (i.e., the relationship between fluorescent intensities and protein copy numbers). Finally, the raw data were translated to protein concentrations based on the calibration curve. Based on this strategy, β-actin of three tumor cell lines of A549, MCF 10A, and Hep G2 was quantified for up to 10,000 single cells. Using a similar platform, the average cell diameter and β-actin concentration of five common cancer cell lines were characterized.[67]

The microfluidic flow cytometry is also capable of measuring multiple cellular proteins when coupling multiple detecting PMTs (Fig. 5).[68] Cellular β-actin, α-tubulin, and β-tubulin in tens of thousands of cells of five tumor cell lines were quantified. In addition, the cellular binding sites of Ras, c-Myc, and p53 from thousands of cells derived from two oral tumor patient samples were accurately characterized. This microfluidic flow cytometry platform allows the quantification of multiple intracellular proteins with a high throughput, which may serve as a useful tool for cell-type classifications.

2.5 Single-cell Western blot

Western blot is a widely used method in cell and molecular biology. It uses specific antibodies to identify proteins in a complex mixture extracted from cells or tissues. However, the sensitivity of traditional western blot is limited and not able to analyze single cell proteins. To address the need for high-specificity protein assays capable of profiling cell-to-cell heterogeneity, a platform named single cell western blot (scWestern) was developed in Herr's Lab.[69] scWestern consists of a microscope slide base and a layer of thin photoactive polyacrylamide (PA) gel patterned with 6720 microwells. The analysis process employs a six-stage assay comprising cell settling, chemical lysis, PA gel electrophoresis (PAGE), UV-initiated protein immobilization onto the gel, diffusion-driven antibody probing and fluorescence imaging. scWestern allows the analysis of up to 48 targets per array (3-plex target quantitation for 16 microwell blocks) in 4 h. It was applied to study immature neuron stem cells and their differentiation over a 6-day period in conditions could yield both astrocytes and neurons. scWestern revealed significant cell-to-cell marker expression heterogeneity over the course of differentiation.

scWestern was also used to study cell-to-cell variations in chemosensitivity and expression of drug-resistance in proteins human glioblastoma cells (Fig. 6).[70] Intact cells seated in microwells were imaged to profile the phenotype, followed by cell lysis and the western analysis. Single glioblastoma cells were treated with the daunomycin, which resulted in both apoptotic and living cells. It is found that living glioblastoma cells upregulated the expression of P-glycoprotein (a multidrug-resistant protein). To support

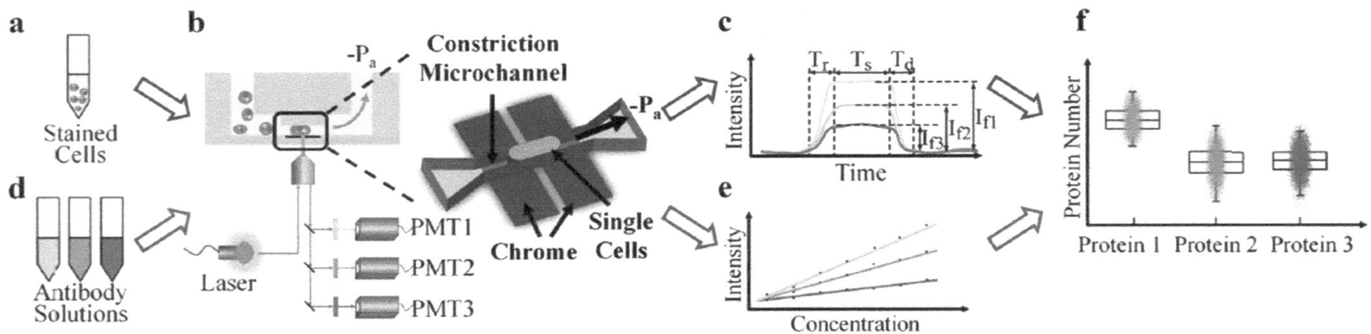

Fig. 5 Schematic of the multiplexed microfluidic flow cytometry. *(Reproduced with permission from Liu, L.; Yang, H.; Men, D.; Wang, M.; Gao, X.; Zhang, T.; Chen, D.; Xue, C.; Wang, Y.; Wang, J.; Chen, J. Development of Microfluidic Platform Capable of High-Throughput Absolute Quantification of Single-Cell Multiple Intracellular Proteins from Tumor Cell Lines and Patient Tumor Samples. Biosens. Bioelectron.* ***2020,*** *155, 112097.)*

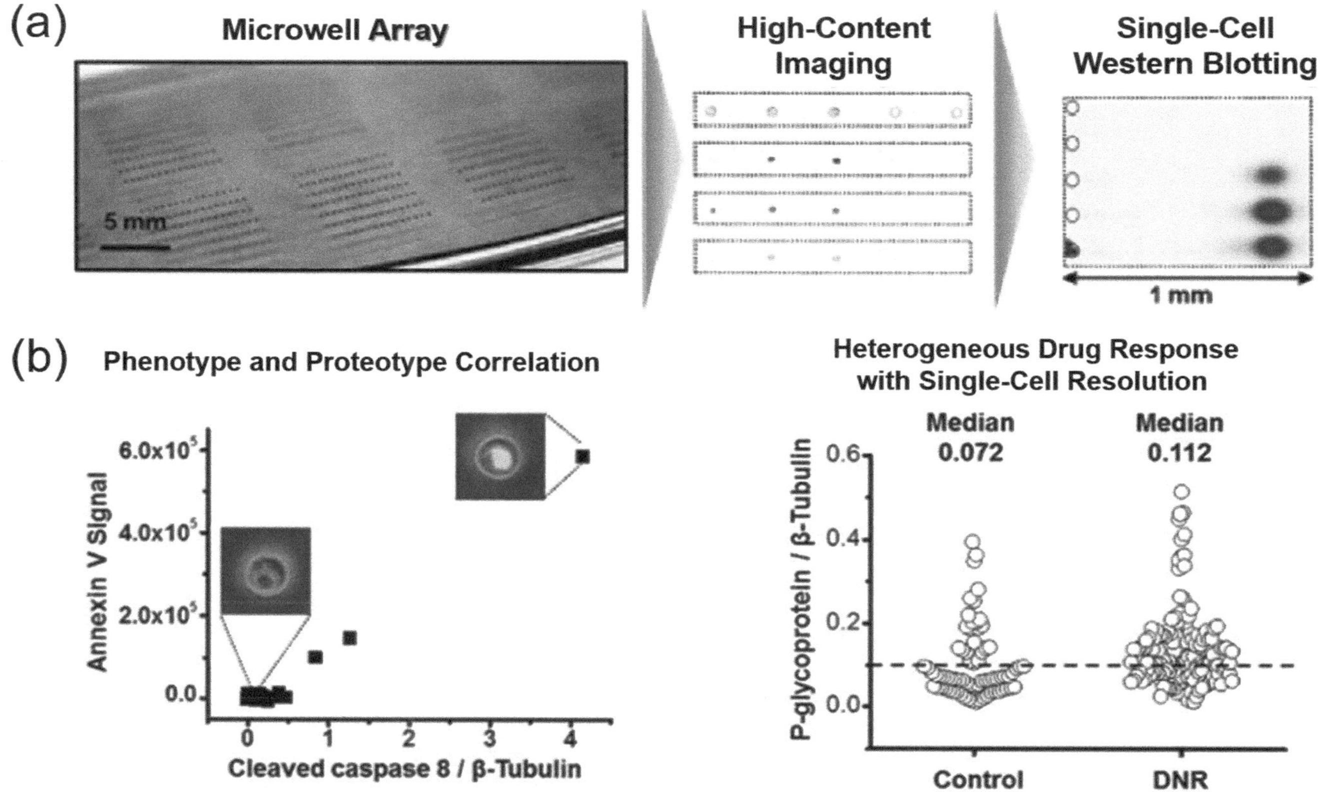

Fig. 6 Protein characterization by single-cell western blot (scWestern). (A) Single-cell phenotype imaging and western blotting workflow. (B) Analysis of daunorubicin-induced single-cell phenotype and targeted proteomics by scWestern. *(Reproduced with permission from Kang, C. C.; Lin, J. M.; Xu, Z.; Kumar, S.; Herr, A. E. Single-Cell Western Blotting After Whole-Cell Imaging to Assess Cancer Chemotherapeutic Response. Anal. Chem.* ***2014,*** *86 (20), 10429–10436.)*

the analyses of low starting cell numbers from precious samples, lab-on-a-disc scWestern platform was proposed.[71] As few as 200 cells could be processed with 75% cell settling efficiencies based on the centrifugation force. This assay achieved a limit of detection of 27,000 molecules per cell, which represents a sensitivity suitable for the analysis of ~50% of the mammalian proteome.[72] CTCs are rare cells circulating in blood at very low occurrence (~1–500 cells per 7.5 mL of blood).[73] To facilitate the protein analysis of rare CTCs, the scWestern assay was integrated with a label-free CTC enriching device based on cell size and deformability.[74] The rare-cell scWestern assay characterized 12 unique protein targets in single cells from three cancer subtypes. Additionally, patient-derived CTCs were found demonstrating a distinct biophysical phenotype. The scWestern method was also applied to analyze the proteins in adherent cells.[22] After overnight single-cell culture, in situ scWestern was conducted in the microwell followed cell lysis. Cell-to-cell heterogeneity in stress responses were found in the cultured HeLa cells. It is anticipated that scWestern-based approaches will open up new possibilities for SCP characterization.

3 Substrate-based characterization of SCP by microfluidic biosensor

Enzymes such as proteases and kinases are main functional proteins in cells.[75] Cellular biochemical processes involve the participation of many enzymes. Characterization of single cell enzymatic kinetics provides detailed information about variability in relevant biochemical reactions and benefits the deep understanding of signaling pathways.[76,77] Owing to the ultralow abundance of the enzymes, conventional methods such as enzyme-linked immunosorbent assay (ELISA) in a well-plate can only determine the average activity of a cell population and is unable to measure the enzyme activity at a single-cell level.[27] Microfluidics-based platforms allows the confinement of single cells or cell lysate in a minimal volume and thus the enzyme can achieve a high concentration even with low quantity.[78] In this section, we review substrate-based single cell enzyme characterization by microfluidic biosensors. Substrate-based assay can directly reflect the enzyme activity, and possesses high sensitivity since a single enzyme turns over up to 1000 substrate molecules to achieve an amplified signal.[79] Herein we mainly summarize the advances in biosensor-based characterization of single cell proteases and kinases, which have many fluorogenic substrates commercially available.

Essential cellular functions are dynamically regulated within living cells.[5, 21, 27] In comparison with immunostaining or ELISA, substrate-based assay is fast and sensitive and facilitates the dynamic detection of enzyme activity.[80, 81] In an early work, Di Carlo et al. reported an array-based single cell trapping device and applied it to determine the carboxylesterase activity in three different cell types of HeLa, 293T, Jurkat cells.[76] By using calcein-AM as the substrate, different carboxylesterase concentration distributions were found within the three human cell lines. The responses of the three cell lines also varied after applying the nonspecific inhibitor nordihydroguaiaretic acid (NDGA). β-galactosidase (β-gal) is an important enzyme for organisms as it plays a key role in energy production by breaking lactose to galactose and glucose.[82] A droplet-based micro-optical system was developed for screening β-gal activity in single *Escherichia coli*.[83] Nonfluorescent substrate fluorescein di-β-D-galactopyranoside (FDG) was cleaved to produce the free fluorescein and generated bright fluorescence signal in the cell-containing droplets. The platform is able to measure over 100 channels equipped with microlenses and analyze the data by image processing.

Matrix metalloproteinases (MMPs) are a family of zinc-dependent enzymes that have more than 20 different members.[84] MMPs can break down extracellular matrix (ECM) proteins such as collagen, elastin, and gelatin, and participate in diverse biological and pathological processes. To measure MMPs secreted from single PC9 cells, a jetting droplet platform was proposed.[80] A fluorescence resonance energy transfer (FRET)-based polypeptide substrate was used to measure the enzymatic activity in individual droplets. To enrich the single-cell encapsulated droplets, a deterministic lateral displacement (DLD) sorting channel was integrated into the system. Single cell encapsulation efficiency achieved up to 80% at a frequency of greater than 5 kHz. The results revealed cell-to-cell heterogeneity in MMPs secretion and the responses to the drug doxycycline. It is hypothesized circulating tumor cells (CTCs) may employ MMPs to assist the invasion and metastasis.[81] To address this issue, Dhar et al. reported an integrated microfluidic system that can concentrate CTCs by $>$100,000-folds from 1 mL whole blood and then isolate it into ~50,000 2-nL drops containing reagents for MMPs assay. It is found that CTCs demonstrated higher MMPs activity than leukocytes from the prostate cancer patient (average CTC/leukocyte MMP activity ratio, 2.6 ± 1.5). These results suggest that MMPs might be highly associated with cancer invasion. Interestingly, many platelets isolated from clinical samples secreted more MMPs than healthy samples, which indicate that cellular components of blood other than CTCs may also contain important information for the elucidation of metastasis mechanism. The earlier mentioned biosensors are based on droplets. They show merits of high throughput and good sensitivity for analyzing MMP of single suspended cells, but have limitations when profiling adherent cells that require a substrate for attaching. To facilitate such studies, we previously reported a virtual-droplet system named single cell protease sensor (SCPS).[85] SCPS achieved a single-cell capture efficiency of >90% in 648 chambers. Different from typical droplet platforms that encapsulate cells during droplet formation, SCPS captures single cells in advance, followed by the injection of oil to isolate each chamber to form a virtual-droplet array (Fig. 7A). The trapped cells were not totally surrounded by water-in-oil droplets, but had the device substrate to support the adhesion. The MMP9 profile of five cell lines, including two high-invasive (MDA-MB-231 and SUM159), two low-invasive (MCF-7 and SK-BR-3), and one normal (MCF-10A) breast cell lines, were investigated in SCPS. All the four cancer cell lines demonstrated significant cell-to-cell heterogeneity in MMP9 secretion and their MMP9 levels are much higher than the normal cell line. It is worth to be noted that individual adhered MDA-MB-231 cells were found secreting relatively more MMP9 than suspended cells. Additionally, the cells producing high MMP9 and low MMP9 were retrieved from the device for downstream mRNA-sequence. The results revealed that some pathways associated with cell migration and cancer invasion were enriched in high-MMP9 cells, which further proved the association of MMPs with cancer metastasis.

Multiplexed profiling of enzymes secreted from individual native immune cells could reveal the underlying mechanism of immune signaling, inflammation, and cytotoxic activity.[27, 37] Using a fully integrated system, Zeming et al. completed a series of tasks from leukocyte extraction to real-time readout in 60 min (Fig. 7B).[86] They studied the leukocytes of acute heart failure (AHF) patients at different states, and profiled the secretion in neutrophil elastase, granzyme B, and metalloproteinase. The results suggest patient-to-patient heterogeneous in enzymatic signatures and a reduction in enzymatic secretion in samples from patients recovering from AHF.

ADAMs (a disintegrin and metalloproteinases) are membrane-anchored enzymes possessing both a potential adhesion domain and a potential protease domain.[87] They convert nearby membrane-bound proteins into soluble bioactive mediators. To study the secretory ADAMs from individual leukocytes, a droplet platform was established by Jing et al.[88] The naive and phorbol 12-myristate 13-acetate (PMA)-stimulated leukocytes demonstrated

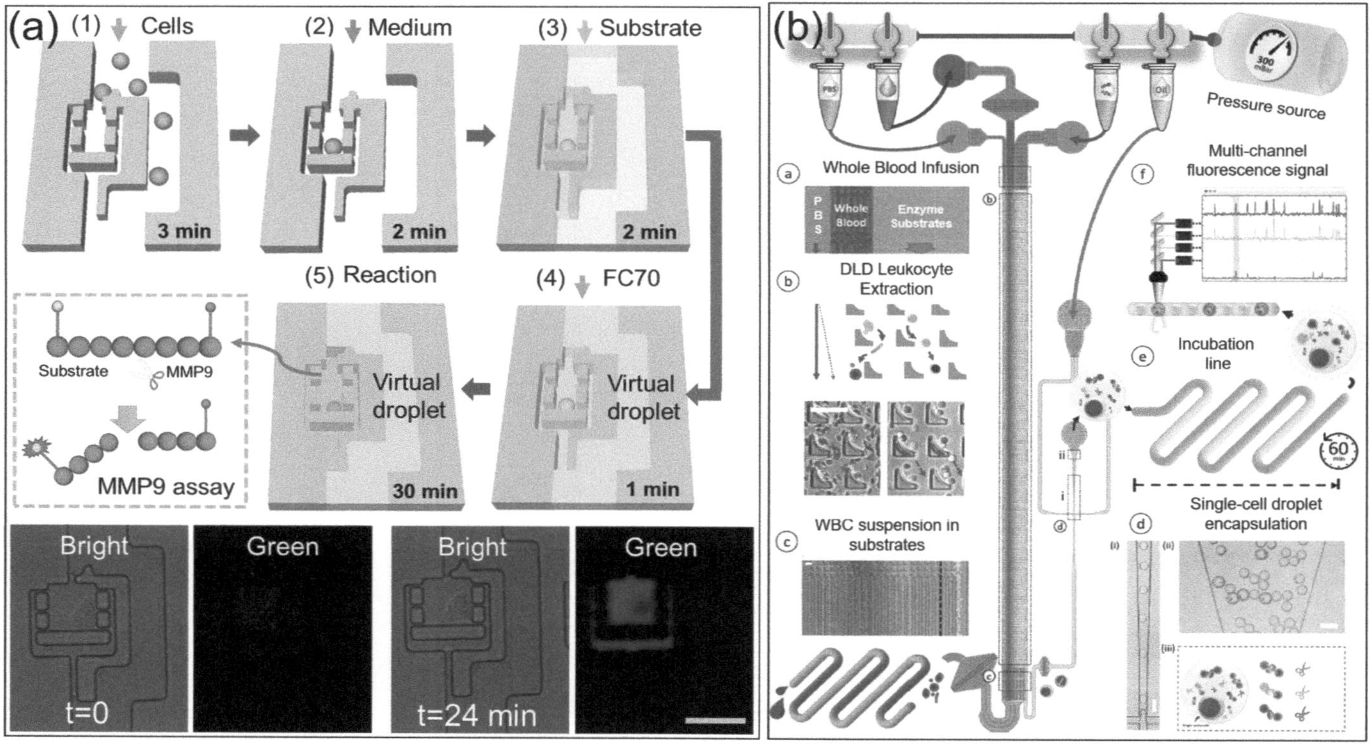

Fig. 7 Substrate-based single cell protease analysis. (A) Workflow of the virtual-droplet platform and on-chip monitoring of MMP9 secretion. (B) Schematics of the integrated microfluidic platform with the DLD section and single-cell droplet assay. *((A) Reproduced with permission from Li, Y.; Zhang, P.; Li, T.; Hu, R.; Zhu, J.; He, T.; Yang, Y.; Liu, M. A Virtual-Droplet System for Sensing MMP9 Activity of Single Suspended and Adhered Cancer Cells. Sens. Actuators B Chem.* ***2020,*** *308, 127749. (B) Reproduced with permission from Zeming, K. K.; Lu, R.; Woo, K. L.; Sun, G.; Quek, K. Y.; Cheow, L. F.; Chen, C. H.; Han, J.; Lim, S. L. Multiplexed Single-Cell Leukocyte Enzymatic Secretion Profiling from Whole Blood Reveals Patient-Specific Immune Signature. Anal. Chem.* ***2021,*** *93 (10), 4374–4382.)*

significant differences in ADAMs activity. It is found that PMA-treated leukocytes showed elevated ADAMs and reduced cellular heterogeneity in protease secretion. To characterize multiple enzymes including ADAM8/10/17 and MMP2/3/9 released by single cells, a droplet-based microfluidic flow cytometer (droplet-μFC) system was developed by combining an automatic optical sensor and a computational analysis method.[89] The authors characterized the proteolytic activities of different cell lines, including tumor stromal components and of cancer cell lines with different aggressiveness. A database library was also established that allows the determination of different cell types in a tumor based on cell mapping. Accordingly, about 0.5 million cells from a patient-derived tumor were analyzed in 2 h. This platform can rapidly analyze the tumor status based on single cell proteolytic activities, which could potentially serve as a useful tool for onsite diagnostics and precision medicine.

Kinases are enzymes that catalyze the transfer of a phosphate group from phosphate-donating molecules to specific substrates.[90] Kinase regulates a majority of cell signaling pathways in response to stimuli and plays a vital role in the determination of cell fate. To measure cellular kinase activity levels, a microfluidic-based strategy was established by combining a nanoliter microwell device and an electrokinetic concentration chip.[91] Cells were isolated and grown in the microwell, and then ultrasonically lysed to let the released intracellular kinases catalyze the phosphorylation of peptide substrates preadded. Three kinases of PKA, Akt, and MK2 were studied. The electrokinetic concentration chip was used to concentrate and separate substrate peptides, enabling simultaneous measurements of the activity of the three kinases. This platform could be applied to study cellular signaling network between different pathways and provide insights on cell regulatory functions.

4 Chip-MS for SCP characterization

Mass spectrometry (MS) is the most effective method in proteome analysis, including peptide mapping, posttranslational modification, protein-protein interactions, and others.[19] Though MS has been applied for SCP analysis,[18, 92, 93] it still faces many technical challenges such as the difficulty in single cell sampling.[26, 94] Microfluidic devices have the ability to handle single cells in parallel and are suitable to provide an interface with MS for high-throughput SCA.[95] In 1997, Karger et al.[96] and Ramsey et al.[97] proposed the concept of microchip-MS (Chip-MS) for biochemical analysis.

The development of Chip-MS technology has greatly accelerated the identification of proteins, now serving as an important detection method for SCM characterization.[29, 98]

An early work was demonstrated by Mellors et al.[99] They developed a continuous-flow microdevice for sampling single cells, followed by the analysis using capillary electrophoresis (CE) and electrospray ionization mass spectrometry (ESI-MS). Human erythrocytes were analyzed in this study, with a throughput of approximately 12 cells per minute. When cells flowed through the device, the dissociated heme group and the α/β subunits of hemoglobin from individual erythrocytes were measured. In 2012, Ros's group presented a method with the combination of a microdevice and a matrix-assisted laser desorption ionization time-of-flight mass spectrometer (MALDI-TOF-MS), which demonstrated a detection limit of 10^6 insulin molecules, approaching a sensitivity for detecting single cell proteins.[100] In 2016, this group reported a quantitative mass spectrometric approach combined with an improved microdevice, reaching the detection sensitivity of high abundant proteins in single cells.[94] With the assist of a series of chambers and valves, on-chip protein digestion and labeling were realized, which also facilitated the absolute quantification of targeted proteins. They quantitatively evaluated the number of Bcl-2 molecules.

To minimize sample loss during single-cell protein analysis, Fang's group developed an LC-MS/MS analysis system combined with a nanoliter-scale oil-air-droplet (OAD) chip and a self-aligning monolithic (SAM) device (Fig. 8A), which achieved multistep complex sample pretreatment and high efficiency sample injection.[101] The system's performance was evaluated with the proteomic analysis of various numbers of HeLa cells. A sum of triplicate runs for 100 ± 10, 50 ± 5, 10, and 1 HeLa cell (s) identified 1360, 612, 192, and 51 proteins, respectively. Additionally, the pipeline was applied to analyze one single mouse oocyte and 355 proteins (summing triplicate measurements) were identified. Recently, Zhu et al. described an integrated platform to measure the proteome in CTCs.[102] Spiked CTCs were enriched from whole blood, lysed and prepared using a nanodroplet-based processing platform (nanoPOTS),[98, 103] and finally analyzed by ultrasensitive nano-LC-MS/MS. The strategy was able to identify an average of 164 and 607 protein groups for 1 and 5 LNCaP cells, respectively. Moreover, a panel of prostate cancer-specific proteins were identified and quantified to differentiate spiked CTCs and white blood cells. By using the similar platform, Zhu et al. introduced tandem mass tag (TMT) isobaric labeling to significantly improve the proteome coverage for single mammalian cells.[104] They

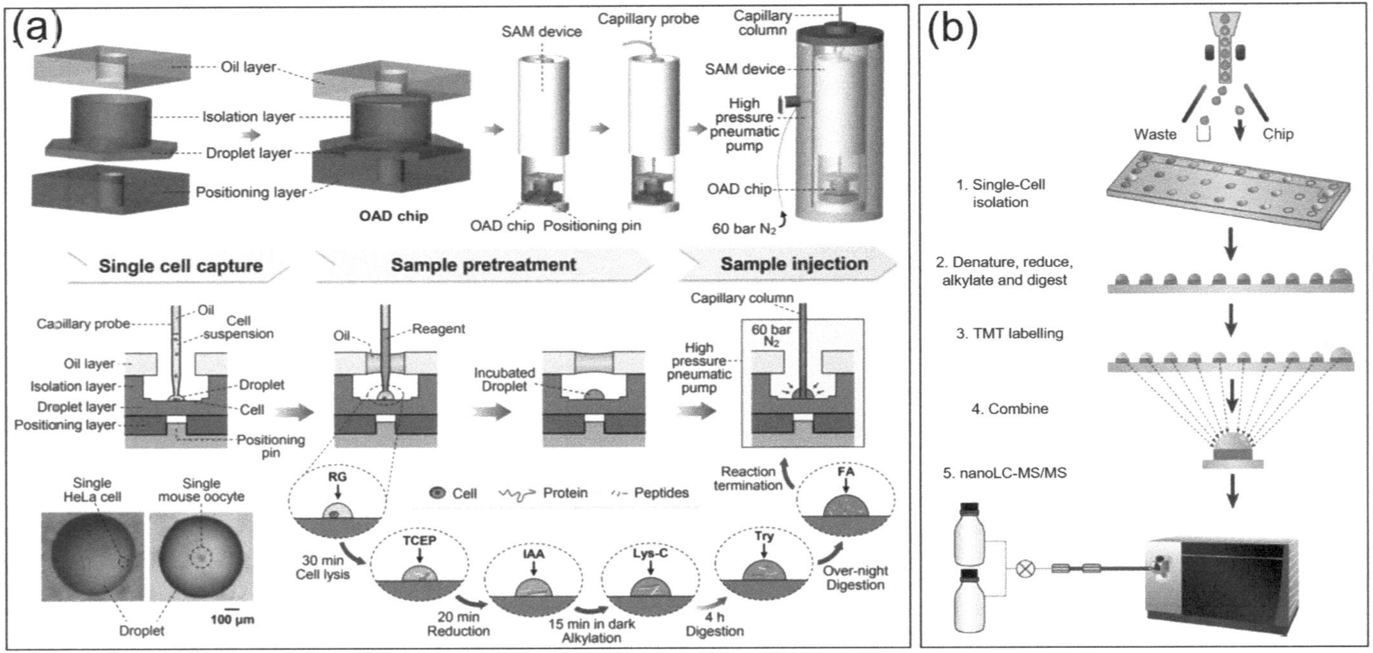

Fig. 8 Integration of microfluidic chip with MS for SCP. (A) Scheme of the oil-air-droplet chip setup and operation of the self-aligning monolithic device. (B) Workflow of the nano-POTS-TMT-based SCP platform. *((A) Reproduced with permission from Li, Z. Y.; Huang, M.; Wang, X. K.; Zhu, Y.; Li, J. S.; Wong, C. C. L.; Fang, Q. Nanoliter-Scale Oil-Air-Droplet Chip-Based Single Cell Proteomic Analysis. Anal. Chem.* ***2018,*** *90 (8), 5430–5438. (B) Reproduced with permission from Dou, M.; Clair, G.; Tsai, C. F.; Xu, K.; Chrisler, W. B.; Sontag, R. L.; Zhao, R.; Moore, R. J.; Liu, T.; Pasa-Tolic, L.; Smith, R. D.; Shi, T.; Adkins, J. N.; Qian, W. J.; Kelly, R. T.; Ansong, C.; Zhu, Y. High-Throughput Single Cell Proteomics Enabled by Multiplex Isobaric Labeling in a Nanodroplet Sample Preparation Platform. Anal. Chem.* ***2019,*** *91 (20), 13119–13127.)*

profiled the proteome in 72 single cells from three types of cells (epithelial, immune, and endothelial cells), and over 2300 proteins were identified.

In summary, Chip-MS has been successfully applied for SCP characterization. We believe that more and more such platforms will be developed with the advance of microfluidics and MS techniques.

5 Commercial microfluidic-based biosensing platform for SCP characterization

Currently, most of microfluidic devices are built in different laboratories. The operation of these devices is not standardized, which renders it difficult to be widely used for SCP characterization in the research community, especially for those biologists without experiences in handling microdevices. In comparison with the custom-designed devices, commercialized products are relatively stable and requiring minimal training to use. C1 system from Fluidigm is an automated microfluidic system that can rapidly and reliably isolate, process, and profile up to 96 individual cells for downstream analysis.[105] Initially, C1 is designated for genome or transcriptome analysis. For example, mRNA of the trapped cells can be extracted, reverse transcribed, preamplified, and detected via qPCR or sequencing.[106] Recently, this platform has been implemented for analyzing single cell proteome by integrating proximity extension assays (PEA).[107] PEA uses a pair of antibodies conjugated with a pair of single-stranded DNA oligonucleotides with complementary 3′ ends.[64] Hence, it is possible to simultaneously and specifically detect large sets of protein and RNA targets for individual cells. This approach was applied to study the effects of a therapeutic agent BMP4 on early-passage glioblastoma cells.[107] Another commercial platform for SCP is IsoPlexis, which utilize the antibody barcoding technology to characterize cellular secretome and proteome heterogeneity within cell populations at different timepoints, especially useful for studying the phenotypes of cancer and immune cells. IsoPlexis was implemented to study drug resistance in melanoma based on single-cell proteomics and metabolomics analysis.[32] Recently, MissionBio and BioLegend collaborated to establish the Tapestri platform for multiomic analysis. This system is able to simultaneously measure up to 1000 DNA and 45 protein targets in thousands of single cells. Over 20,000 single cells from six AML patient samples were analyzed on Tapestri with a custom DNA panel and a custom oligo-conjugated antibody panel and the data were processed with a Pipeline named Tapestri Insights.[2] This work demonstrated the use of a commercialized toolbox for the investigation of genotype and phenotype correlation across thousands of cells. Although there are now few commercialized platforms capable of performing SCP analysis, we anticipate more and more products to be launched with the advance of microfluidic technologies and the promotion of biomedical demands.

6 Conclusion and perspectives

SCM analysis enables the characterization of protein activity and function at cellular level. With cell-matched size, minimal sample consumption, and controllability, microfluidics-based biosensors offer unique advantages for SCM characterization. As summarized in this chapter, these novel and diverse biosensing platforms have enabled important discoveries of cellular development and heterogeneity. It improves our understanding of the cellular proteomic patterns and relevant biological and pathological processes, and contributes to both fundamental research and precision medicine.

Notwithstanding the exciting advances, there are still nontrivial challenges to be addressed for SCM characterization. First, the proteins that can be measured in a microfluidic-based biosensor are still limited. Proteins are in a complex interaction to function within a cell. However, most of current biosensors can only quantify <5 proteins, though SCBCs can dynamically measure ~ 40 proteins at a time.[38, 43] The PLA- or PEA-based approach can detect multiple proteins by converting protein to DNA sequence through oligonucleotide-conjugated antibody.[63, 64] But this method is still hard to achieve a throughput higher than 100 protein targets. To better elucidate direct and indirect protein interactions and signaling mechanisms, it is necessary to introduce new labeling techniques to cover a broader proteome. The combination of micro-devices with MS could be a good way to detect more proteins in a single cell, but the sampling efficiency is still too low due to the offline integration.[19, 29, 35] It is expected that a fully integrated system can automatically complete the sample preparation in a micro-device and deliver the contents to the MS analyzer for downstream SCM analysis. Next, it is now still difficult to perform multiomics assays for the same cell in current biosensors.[108] Though a few studies have been performed to combine proteome and transcriptome for the elucidation of single cell heterogeneity.[1, 62] More efforts are expected to put into this aspect. Integration of information from genomics, epigenomics, transcriptomics, proteomics, and metabolomics will enable a more accurate and comprehensive characterization of a cell state and function. It will be very meaningful to reveal the mechanism of single-cell level molecular translation to cellular phenotypic properties. Additionally, retaining spatial information during single cell sampling and combining multiomics data from different organisms of human body will build a comprehensive human cell atlas, which should impact many fields such as biological research and precision medicine.[12, 109]

Moreover, it lacks data analysis tools for analyzing the huge SCM data. It could take several days or weeks to manually analyze the data and perform in-depth comparisons.[19, 26, 29] In consideration of this problem, it is necessary to establish collaborations between biochemists and computer scientists to develop some pipelines to automatically analyze the produced data (e.g., develop deep learning models).[110] A data processing pipeline would also benefit data sharing between the community. Finally, there are very limited commercialized biosensors for SCM analysis. Though many biosensing platforms have been built in different labs, most of them require extensive expertise and too much manual work. To become a widespread technology in clinical applications, the system should be reproducible and robust. A potential strategy to resolve such problem is to improve the device automation and simplify the operation protocol.

Despite all these challenges, the future of microfluidics-based biosensor for SCP characterization is promising. With the driving force of clinical applications and the deep collaboration among chemists, biologists, and engineers, this field will flourish at an ever-faster rate and eventually benefits human health and disease curing.

Acknowledgment

The authors acknowledge the financial supports from National Key R&D Program of China (2018YFA0704000, 2018YFE0202300), National Natural Science Foundation of China (21904139, 21991080), and Chinese Academy of Sciences (Y9Y1041001, YJKYYQ20170026).

References

1. Stoeckius, M.; Hafemeister, C.; Stephenson, W.; Houck-Loomis, B.; Chattopadhyay, P. K.; Swerdlow, H.; Satija, R.; Smibert, P. Simultaneous Epitope and Transcriptome Measurement in Single Cells. *Nat. Methods* **2017**, *14* (9), 865–868.
2. Miles, L. A.; Bowman, R. L.; Merlinsky, T. R.; Csete, I. S.; Ooi, A. T.; Durruthy-Durruthy, R.; Bowman, M.; Famulare, C.; Patel, M. A.; Mendez, P.; Ainali, C.; Demaree, B.; Delley, C. L.; Abate, A. R.; Manivannan, M.; Sahu, S.; Goldberg, A. D.; Bolton, K. L.; Zehir, A.; Rampal, R.; Carroll, M. P.; Meyer, S. E.; Viny, A. D.; Levine, R. L. Single-Cell Mutation Analysis of Clonal Evolution in Myeloid Malignancies. *Nature* **2020**, *587* (7834), 477–482.

3. Slavov, N. Unpicking the Proteome in Single Cells. *Science* **2020**, *367* (6477), 512–513.
4. Yu, J.; Zhou, J.; Sutherland, A.; Wei, W.; Shin, Y. S.; Xue, M.; Heath, J. R. Microfluidics-Based Single-Cell Functional Proteomics for Fundamental and Applied Biomedical Applications. *Annu. Rev. Anal. Chem.* **2014**, 7, 275–295.
5. Lu, Y.; Yang, L.; Wei, W.; Shi, Q. Microchip-Based Single-Cell Functional Proteomics for Biomedical Applications. *Lab Chip* **2017**, *17* (7), 1250–1263.
6. Lombard-Banek, C.; Moody, S. A.; Nemes, P. Single-Cell Mass Spectrometry for Discovery Proteomics: Quantifying Translational Cell Heterogeneity in the 16-Cell Frog (Xenopus) Embryo. *Angew. Chem. Int. Ed. Engl.* **2016**, *55* (7), 2454–2458.
7. Tentori, A. M.; Yamauchi, K. A.; Herr, A. E. Detection of Isoforms Differing by a Single Charge Unit in Individual Cells. *Angew. Chem. Int. Ed. Engl.* **2016**, *55* (40), 12431–12435.
8. Heath, J. R.; Ribas, A.; Mischel, P. S. Single-Cell Analysis Tools for Drug Discovery and Development. *Nat. Rev. Drug Discov.* **2016**, *15* (3), 204–216.
9. Specht, H.; Slavov, N. Transformative Opportunities for Single-Cell Proteomics. *J. Proteome Res.* **2018**, *17* (8), 2565–2571.
10. Labib, M.; Wang, Z.; Ahmed, S. U.; Mohamadi, R. M.; Duong, B.; Green, B.; Sargent, E. H.; Kelley, S. O. Tracking the Expression of Therapeutic Protein Targets in Rare Cells by Antibody-Mediated Nanoparticle Labelling and Magnetic Sorting. *Nat. Biomed. Eng.* **2021**, *5* (1), 41–52.
11. Labib, M.; Kelley, S. O. Single-Cell Analysis Targeting the Proteome. *Nat. Rev. Chem.* **2020**, *4* (3), 143–158.
12. Uhlen, M.; Fagerberg, L.; Hallstrom, B. M.; Lindskog, C.; Oksvold, P.; Mardinoglu, A.; Sivertsson, A.; Kampf, C.; Sjostedt, E.; Asplund, A.; Olsson, I.; Edlund, K.; Lundberg, E.; Navani, S.; Szigyarto, C. A.; Odeberg, J.; Djureinovic, D.; Takanen, J. O.; Hober, S.; Alm, T.; Edqvist, P. H.; Berling, H.; Tegel, H.; Mulder, J.; Rockberg, J.; Nilsson, P.; Schwenk, J. M.; Hamsten, M.; von Feilitzen, K.; Forsberg, M.; Persson, L.; Johansson, F.; Zwahlen, M.; von Heijne, G.; Nielsen, J.; Ponten, F. Proteomics. Tissue-Based Map of the Human Proteome. *Science* **2015**, *347* (6220), 1260419.
13. Wilkins, M. R.; Pasquali, C.; Appel, R. D.; Ou, K.; Golaz, O.; Sanchez, J. C.; Yan, J. X.; Gooley, A. A.; Hughes, G.; HumpherySmith, I.; Williams, K. L.; Hochstrasser, D. F. From Proteins to Proteomes: Large Scale Protein Identification by Two-Dimensional Electrophoresis and Amino Acid Analysis. *Biotechnology* **1996**, *14* (1), 61–65.
14. Zhang, L.; Vertes, A. Single-Cell Mass Spectrometry Approaches to Explore Cellular Heterogeneity. *Angew. Chem. Int. Ed. Engl.* **2018**, *57* (17), 4466–4477.
15. Lederer, A. R.; La Manno, G. The Emergence and Promise of Single-Cell Temporal-Omics Approaches. *Curr. Opin. Biotechnol.* **2020**, *63*, 70–78.
16. Kelly, R. T. Single-Cell Proteomics: Progress and Prospects. *Mol. Cell. Proteomics* **2020**, *19* (11), 1739–1748.
17. Liu, L.; Chen, D.; Wang, J.; Chen, J. Advances of Single-Cell Protein Analysis. *Cells* **2020**, *9* (5).
18. Pham, T.; Tyagi, A.; Wang, Y. S.; Guo, J. Single-Cell Proteomic Analysis. *Wiley Interdiscip. Rev. Syst. Biol. Med.* **2021**, *13* (1), e1503.
19. Timp, W.; Timp, G. Beyond Mass Spectrometry, the Next Step in Proteomics. *Sci. Adv.* **2020**, *6* (2).
20. Milo, R.; Jorgensen, P.; Moran, U.; Weber, G.; Springer, M. BioNumbers—The Database of Key Numbers in Molecular and Cell Biology. *Nucleic Acids Res.* **2010**, *38* (Database issue), D750–D753.
21. Chen, P.; Chen, D.; Li, S.; Ou, X.; Liu, B.-F. Microfluidics Towards Single Cell Resolution Protein Analysis. *Trends Anal. Chem.* **2019**, *117*, 2–12.
22. Liu, Y.; Chen, X.; Zhang, Y.; Liu, J. Advancing Single-Cell Proteomics and Metabolomics with Microfluidic Technologies. *Analyst* **2019**, *144* (3), 846–858.

23. Lahav, G.; Rosenfeld, N.; Sigal, A.; Geva-Zatorsky, N.; Levine, A. J.; Elowitz, M. B.; Alon, U. Dynamics of the p53-Mdm2 Feedback Loop in Individual Cells. *Nat. Genet.* **2004**, *36* (2), 147–150.
24. Irish, J. M.; Myklebust, J. H.; Alizadeh, A. A.; Houot, R.; Sharman, J. P.; Czerwinski, D. K.; Nolan, G. P.; Levy, R. B-Cell Signaling Networks Reveal a Negative Prognostic Human Lymphoma Cell Subset that Emerges During Tumor Progression. *Proc. Natl. Acad. Sci. U. S. A.* **2010**, *107* (29), 12747–12754.
25. Deng, Y.; Finck, A.; Fan, R. Single-Cell Omics Analyses Enabled by Microchip Technologies. *Annu. Rev. Biomed. Eng.* **2019**, *21*, 365–393.
26. Spitzer, M. H.; Nolan, G. P. Mass Cytometry: Single Cells, Many Features. *Cell* **2016**, *165* (4), 780–791.
27. Chen, Z.; Chen, J. J.; Fan, R. Single-Cell Protein Secretion Detection and Profiling. *Annu. Rev. Anal. Chem.* **2019**, *12* (1), 431–449.
28. Jackson, H. W.; Fischer, J. R.; Zanotelli, V. R. T.; Ali, H. R.; Mechera, R.; Soysal, S. D.; Moch, H.; Muenst, S.; Varga, Z.; Weber, W. P.; Bodenmiller, B. The Single-Cell Pathology Landscape of Breast Cancer. *Nature* **2020**, *578* (7796), 615–620.
29. Slavov, N. Single-Cell Protein Analysis by Mass Spectrometry. *Curr. Opin. Chem. Biol.* **2021**, *60*, 1–9.
30. Yang, Y.; Chen, Y.; Tang, H.; Zong, N.; Jiang, X. Microfluidics for Biomedical Analysis. *Small Methods* **2020**, *4* (4), 1900451.
31. Li, Y.; Jang, J. H.; Wang, C.; He, B.; Zhang, K.; Zhang, P.; Vu, T.; Qin, L. Microfluidics Cell Loading-Dock System: Ordered Cellular Array for Dynamic Lymphocyte-Communication Study. *Adv. Biosyst.* **2017**, *1* (10), e1700085.
32. Su, Y.; Ko, M. E.; Cheng, H.; Zhu, R.; Xue, M.; Wang, J.; Lee, J. W.; Frankiw, L.; Xu, A.; Wong, S.; Robert, L.; Takata, K.; Yuan, D.; Lu, Y.; Huang, S.; Ribas, A.; Levine, R.; Nolan, G. P.; Wei, W.; Plevritis, S. K.; Li, G.; Baltimore, D.; Heath, J. R. Multi-Omic Single-Cell Snapshots Reveal Multiple Independent Trajectories to Drug Tolerance in a Melanoma Cell Line. *Nat. Commun.* **2020**, *11* (1), 2345.
33. Cohen, L.; Walt, D. R. Single-Molecule Arrays for Protein and Nucleic Acid Analysis. *Annu. Rev. Anal. Chem.* **2017**, *10* (1), 345–363.
34. Jie, M.; Mao, S.; Li, H.; Lin, J.-M. Multi-Channel Microfluidic Chip-Mass Spectrometry Platform for Cell Analysis. *Chin. Chem. Lett.* **2017**, *28* (8), 1625–1630.
35. Yu, X.; He, M.; Chen, B.; Hu, B. Recent Advances in Single-Cell Analysis by Inductively Coupled Plasma-Mass Spectrometry: A Review. *Anal. Chim. Acta* **2020**, *1137*, 191–207.
36. Dincer, C.; Bruch, R.; Costa-Rama, E.; Fernandez-Abedul, M. T.; Merkoci, A.; Manz, A.; Urban, G. A.; Guder, F. Disposable Sensors in Diagnostics, Food, and Environmental Monitoring. *Adv. Mater.* **2019**, *31* (30), e1806739.
37. Chen, Z.; Lu, Y.; Zhang, K. R.; Xiao, Y.; Lu, J.; Fan, R. Multiplexed, Sequential Secretion Analysis of the Same Single Cells Reveals Distinct Effector Response Dynamics Dependent on the Initial Basal State. *Adv. Sci.* **2019**, *6* (9), 11.
38. Shi, Q.; Qin, L.; Wei, W.; Geng, F.; Fan, R.; Shin, Y. S.; Guo, D.; Hood, L.; Mischel, P. S.; Heath, J. R. Single-Cell Proteomic Chip for Profiling Intracellular Signaling Pathways in Single Tumor Cells. *Proc. Natl. Acad. Sci. U. S. A.* **2012**, *109* (2), 419–424.
39. Prakadan, S. M.; Shalek, A. K.; Weitz, D. A. Scaling by Shrinking: Empowering Single-Cell 'Omics' with Microfluidic Devices. *Nat. Rev. Genet.* **2017**, *18* (6), 345–361.
40. Xue, M.; Wei, W.; Su, Y.; Kim, J.; Shin, Y. S.; Mai, W. X.; Nathanson, D. A.; Heath, J. R. Chemical Methods for the Simultaneous Quantitation of Metabolites and Proteins from Single Cells. *J. Am. Chem. Soc.* **2015**, *137* (12), 4066–4069.
41. Lu, Y.; Chen, J. J.; Mu, L.; Xue, Q.; Wu, Y.; Wu, P. H.; Li, J.; Vortmeyer, A. O.; Miller-Jensen, K.; Wirtz, D.; Fan, R. High-Throughput Secretomic Analysis of Single Cells to Assess Functional Cellular Heterogeneity. *Anal. Chem.* **2013**, *85* (4), 2548–2556.

42. Elitas, M.; Brower, K.; Lu, Y.; Chen, J. J.; Fan, R. A Microchip Platform for Interrogating Tumor-Macrophage Paracrine Signaling at the Single-Cell Level. *Lab Chip* **2014**, *14* (18), 3582–3588.
43. Lu, Y.; Xue, Q.; Eisele, M. R.; Sulistijo, E. S.; Brower, K.; Han, L.; Amir el, A. D.; Pe'er, D.; Miller-Jensen, K.; Fan, R. Highly Multiplexed Profiling of Single-Cell Effector Functions Reveals Deep Functional Heterogeneity in Response to Pathogenic Ligands. *Proc. Natl. Acad. Sci. U. S. A.* **2015**, *112* (7), E607–E615.
44. Xu, R.; Rai, A.; Chen, M.; Suwakulsiri, W.; Greening, D. W.; Simpson, R. J. Extracellular Vesicles in Cancer—Implications for Future Improvements in Cancer Care. *Nat. Rev. Clin. Oncol.* **2018**, *15* (10), 617–638.
45. Ji, Y.; Qi, D.; Li, L.; Su, H.; Li, X.; Luo, Y.; Sun, B.; Zhang, F.; Lin, B.; Liu, T.; Lu, Y. Multiplexed Profiling of Single-Cell Extracellular Vesicles Secretion. *Proc. Natl. Acad. Sci. U. S. A.* **2019**, *116* (13), 5979–5984.
46. Rodriguez-Moncayo, R.; Jimenez-Valdes, R. J.; Gonzalez-Suarez, A. M.; Garcia-Cordero, J. L. Integrated Microfluidic Device for Functional Secretory Immunophenotyping of Immune Cells. *ACS Sens.* **2020**, *5* (2), 353–361.
47. Dimov, I. K.; Lu, R.; Lee, E. P.; Seita, J.; Sahoo, D.; Park, S. M.; Weissman, I. L.; Lee, L. P. Discriminating Cellular Heterogeneity Using Microwell-Based RNA Cytometry. *Nat. Commun.* **2014**, *5*, 3451.
48. Torres, A. J.; Hill, A. S.; Love, J. C. Nanowell-Based Immunoassays for Measuring Single-Cell Secretion: Characterization of Transport and Surface Binding. *Anal. Chem.* **2014**, *86* (23), 11562–11569.
49. Love, J. C.; Ronan, J. L.; Grotenbreg, G. M.; van der Veen, A. G.; Ploegh, H. L. A Microengraving Method for Rapid Selection of Single Cells Producing Antigen-Specific Antibodies. *Nat. Biotechnol.* **2006**, *24* (6), 703–707.
50. Han, Q.; Bradshaw, E. M.; Nilsson, B.; Hafler, D. A.; Love, J. C. Multidimensional Analysis of the Frequencies and Rates of Cytokine Secretion from Single Cells by Quantitative Microengraving. *Lab Chip* **2010**, *10* (11), 1391–1400.
51. Yamanaka, Y. J.; Szeto, G. L.; Gierahn, T. M.; Forcier, T. L.; Benedict, K. F.; Brefo, M. S.; Lauffenburger, D. A.; Irvine, D. J.; Love, J. C. Cellular Barcodes for Efficiently Profiling Single-Cell Secretory Responses by Microengraving. *Anal. Chem.* **2012**, *84* (24), 10531–10536.
52. Zhao, P.; Bhowmick, S.; Yu, J.; Wang, J. Highly Multiplexed Single-Cell Protein Profiling with Large-Scale Convertible DNA-Antibody Barcoded Arrays. *Adv. Sci.* **2018**, *5* (9), 1800672.
53. Zhou, Y.; Shao, N.; Bessa de Castro, R.; Zhang, P.; Ma, Y.; Liu, X.; Huang, F.; Wang, R. F.; Qin, L. Evaluation of Single-Cell Cytokine Secretion and Cell-Cell Interactions with a Hierarchical Loading Microwell Chip. *Cell Rep.* **2020**, *31* (4), 107574.
54. Liu, Y.; Singh, A. K. Microfluidic Platforms for Single-Cell Protein Analysis. *J Lab Autom* **2013**, *18* (6), 446–454.
55. Duncombe, T. A.; Tentori, A. M.; Herr, A. E. Microfluidics: Reframing Biological Enquiry. *Nat. Rev. Mol. Cell Biol.* **2015**, *16* (9), 554–567.
56. Brouzes, E.; Medkova, M.; Savenelli, N.; Marran, D.; Twardowski, M.; Hutchison, J. B.; Rothberg, J. M.; Link, D. R.; Perrimon, N.; Samuels, M. L. Droplet Microfluidic Technology for Single-Cell High-Throughput Screening. *Proc. Natl. Acad. Sci. U. S. A.* **2009**, *106* (34), 14195–14200.
57. Agresti, J. J.; Antipov, E.; Abate, A. R.; Ahn, K.; Rowat, A. C.; Baret, J. C.; Marquez, M.; Klibanov, A. M.; Griffiths, A. D.; Weitz, D. A. Ultrahigh-Throughput Screening in Drop-Based Microfluidics for Directed Evolution. *Proc. Natl. Acad. Sci. U. S. A.* **2010**, *107* (9), 4004–4009.
58. Mazutis, L.; Gilbert, J.; Ung, W. L.; Weitz, D. A.; Griffiths, A. D.; Heyman, J. A. Single-Cell Analysis and Sorting Using Droplet-Based Microfluidics. *Nat. Protoc.* **2013**, *8* (5), 870–891.

59. Eyer, K.; Doineau, R. C. L.; Castrillon, C. E.; Briseno-Roa, L.; Menrath, V.; Mottet, G.; England, P.; Godina, A.; Brient-Litzler, E.; Nizak, C.; Jensen, A.; Griffiths, A. D.; Bibette, J.; Bruhns, P.; Baudry, J. Single-Cell Deep Phenotyping of IgG-Secreting Cells for High-Resolution Immune Monitoring. *Nat. Biotechnol.* **2017**, *35* (10), 977–982.
60. Shahi, P.; Kim, S. C.; Haliburton, J. R.; Gartner, Z. J.; Abate, A. R. Abseq: Ultrahigh-Throughput Single Cell Protein Profiling with Droplet Microfluidic Barcoding. *Sci. Rep.* **2017**, 7, 44447.
61. Segaliny, A. I.; Li, G.; Kong, L.; Ren, C.; Chen, X.; Wang, J. K.; Baltimore, D.; Wu, G.; Zhao, W. Functional TCR T Cell Screening Using Single-Cell Droplet Microfluidics. *Lab Chip* **2018**, *18* (24), 3733–3749.
62. Lake, B. B.; Chen, S.; Sos, B. C.; Fan, J.; Kaeser, G. E.; Yung, Y. C.; Duong, T. E.; Gao, D.; Chun, J.; Kharchenko, P. V.; Zhang, K. Integrative Single-Cell Analysis of Transcriptional and Epigenetic States in the Human Adult Brain. *Nat. Biotechnol.* **2018**, *36* (1), 70–80.
63. Albayrak, C.; Jordi, C. A.; Zechner, C.; Lin, J.; Bichsel, C. A.; Khammash, M.; Tay, S. Digital Quantification of Proteins and mRNA in Single Mammalian Cells. *Mol. Cell* **2016**, *61* (6), 914–924.
64. Gullberg, M.; Gustafsdottir, S. M.; Schallmeiner, E.; Jarvius, J.; Bjarnegard, M.; Betsholtz, C.; Landegren, U.; Fredriksson, S. Cytokine Detection by Antibody-Based Proximity Ligation. *Proc. Natl. Acad. Sci. U. S. A.* **2004**, *101* (22), 8420–8424.
65. Fan, B.; Li, X.; Chen, D.; Peng, H.; Wang, J.; Chen, J. Development of Microfluidic Systems Enabling High-Throughput Single-Cell Protein Characterization. *Sensors (Basel)* **2016**, *16* (2), 232.
66. Li, X.; Fan, B.; Cao, S.; Chen, D.; Zhao, X.; Men, D.; Yue, W.; Wang, J.; Chen, J. A Microfluidic Flow Cytometer Enabling Absolute Quantification of Single-Cell Intracellular Proteins. *Lab Chip* **2017**, *17* (18), 3129–3137.
67. Li, X.; Fan, B.; Liu, L.; Chen, D.; Cao, S.; Men, D.; Wang, J.; Chen, J. A Microfluidic Fluorescent Flow Cytometry Capable of Quantifying Cell Sizes and Numbers of Specific Cytosolic Proteins. *Sci. Rep.* **2018**, *8* (1), 14229.
68. Liu, L.; Yang, H.; Men, D.; Wang, M.; Gao, X.; Zhang, T.; Chen, D.; Xue, C.; Wang, Y.; Wang, J.; Chen, J. Development of Microfluidic Platform Capable of High-Throughput Absolute Quantification of Single-Cell Multiple Intracellular Proteins from Tumor Cell Lines and Patient Tumor Samples. *Biosens. Bioelectron.* **2020**, *155*, 112097.
69. Hughes, A. J.; Spelke, D. P.; Xu, Z.; Kang, C.-C.; Schaffer, D.V.; Herr, A. E. Single-Cell Western Blotting. *Nat. Methods* **2014**, *11* (7), 749–755.
70. Kang, C. C.; Lin, J. M.; Xu, Z.; Kumar, S.; Herr, A. E. Single-Cell Western Blotting After Whole-Cell Imaging to Assess Cancer Chemotherapeutic Response. *Anal. Chem.* **2014**, *86* (20), 10429–10436.
71. Kim, J. J.; Sinkala, E.; Herr, A. E. High-Selectivity Cytology Via Lab-on-a-Disc Western Blotting of Individual Cells. *Lab Chip* **2017**, *17* (5), 855–863.
72. Kang, C. C.; Yamauchi, K. A.; Vlassakis, J.; Sinkala, E.; Duncombe, T. A.; Herr, A. E. Single Cell-Resolution Western Blotting. *Nat. Protoc.* **2016**, *11* (8), 1508–1530.
73. Nagrath, S.; Sequist, L.V.; Maheswaran, S.; Bell, D. W.; Irimia, D.; Ulkus, L.; Smith, M. R.; Kwak, E. L.; Digumarthy, S.; Muzikansky, A.; Ryan, P.; Balis, U. J.; Tompkins, R. G.; Haber, D. A.; Toner, M. Isolation of Rare Circulating Tumour Cells in Cancer Patients by Microchip Technology. *Nature* **2007**, *450* (7173), 1235–1239.
74. Sinkala, E.; Sollier-Christen, E.; Renier, C.; Rosas-Canyelles, E.; Che, J.; Heirich, K.; Duncombe, T. A.; Vlassakis, J.; Yamauchi, K. A.; Huang, H.; Jeffrey, S. S.; Herr, A. E. Profiling Protein Expression in Circulating Tumour Cells Using Microfluidic Western Blotting. *Nat. Commun.* **2017**, *8*, 14622.
75. Kovarik, M. L.; Allbritton, N. L. Measuring Enzyme Activity in Single Cells. *Trends Biotechnol.* **2011**, *29* (5), 222–230.

76. Di Carlo, D.; Aghdam, N.; Lee, L. P. Single-Cell Enzyme Concentrations, Kinetics, and Inhibition Analysis Using High-Density Hydrodynamic Cell Isolation Arrays. *Anal. Chem.* **2006**, *78* (14), 4925–4930.
77. Chen, C. H.; Miller, M. A.; Sarkar, A.; Beste, M. T.; Isaacson, K. B.; Lauffenburger, D. A.; Griffith, L. G.; Han, J. Multiplexed Protease Activity Assay for Low-Volume Clinical Samples Using Droplet-Based Microfluidics and its Application to Endometriosis. *J. Am. Chem. Soc.* **2013**, *135* (5), 1645–1648.
78. Yang, L.; George, J.; Wang, J. Deep Profiling of Cellular Heterogeneity by Emerging Single-Cell Proteomic Technologies. *Proteomics* **2020**, *20* (13), e1900226.
79. Smejkal, G. B.; Kakumanu, S. Enzymes and Their Turnover Numbers. *Expert Rev. Proteomics* **2019**, *16* (7), 543–544.
80. Jing, T.; Ramji, R.; Warkiani, M. E.; Han, J.; Lim, C. T.; Chen, C.-H. Jetting Microfluidics with Size-Sorting Capability for Single-Cell Protease Detection. *Biosens. Bioelectron.* **2015**, *66*, 19–23.
81. Dhar, M.; Lam, J. N.; Walser, T.; Dubinett, S. M.; Rettig, M. B.; Di Carlo, D. Functional Profiling of Circulating Tumor Cells with an Integrated Vortex Capture and Single-Cell Protease Activity Assay. *Proc. Natl. Acad. Sci. U. S. A.* **2018**, *115* (40), 9986–9991.
82. Juers, D. H.; Matthews, B. W.; Huber, R. E. LacZ Beta-Galactosidase: Structure and Function of an Enzyme of Historical and Molecular Biological Importance. *Protein Sci.* **2012**, *21* (12), 1792–1807.
83. Lim, J.; Vrignon, J.; Gruner, P.; Karamitros, C. S.; Konrad, M.; Baret, J. C. Ultra-High Throughput Detection of Single Cell Beta-Galactosidase Activity in Droplets Using Micro-Optical Lens Array. *Appl. Phys. Lett.* **2013**, *103* (20), 203704.
84. Gonzalez-Avila, G.; Sommer, B.; Mendoza-Posada, D. A.; Ramos, C.; Garcia-Hernandez, A. A.; Falfan-Valencia, R. Matrix Metalloproteinases Participation in the Metastatic Process and Their Diagnostic and Therapeutic Applications in Cancer. *Crit. Rev. Oncol. Hematol.* **2019**, *137*, 57–83.
85. Li, Y.; Zhang, P.; Li, T.; Hu, R.; Zhu, J.; He, T.; Yang, Y.; Liu, M. A Virtual-Droplet System for Sensing MMP9 Activity of Single Suspended and Adhered Cancer Cells. *Sens. Actuators B Chem.* **2020**, *308*, 127749.
86. Zeming, K. K.; Lu, R.; Woo, K. L.; Sun, G.; Quek, K. Y.; Cheow, L. F.; Chen, C. H.; Han, J.; Lim, S. L. Multiplexed Single-Cell Leukocyte Enzymatic Secretion Profiling from Whole Blood Reveals Patient-Specific Immune Signature. *Anal. Chem.* **2021**, *93* (10), 4374–4382.
87. Lambrecht, B. N.; Vanderkerken, M.; Hammad, H. The Emerging Role of ADAM Metalloproteinases in Immunity. *Nat. Rev. Immunol.* **2018**, *18* (12), 745–758.
88. Jing, T.; Lai, Z.; Wu, L.; Han, J.; Lim, C. T.; Chen, C. H. Single Cell Analysis of Leukocyte Protease Activity Using Integrated Continuous-Flow Microfluidics. *Anal. Chem.* **2016**, *88* (23), 11750–11757.
89. Ng, E. X.; Sun, G.; Wei, S. C.; Miller, M. A.; DasGupta, R.; Lam, P. Y. P.; Chen, C. H. Ultrafast Single-Cell Level Enzymatic Tumor Profiling. *Anal. Chem.* **2019**, *91* (2), 1277–1285.
90. Roskoski, R., Jr. Properties of FDA-Approved Small Molecule Protein Kinase Inhibitors: A 2020 Update. *Pharmacol. Res.* **2020**, *152*, 104609.
91. Cheow, L. F.; Sarkar, A.; Kolitz, S.; Lauffenburger, D.; Han, J. Detecting Kinase Activities from Single Cell Lysate Using Concentration-Enhanced Mobility Shift Assay. *Anal. Chem.* **2014**, *86* (15), 7455–7462.
92. Budnik, B.; Levy, E.; Harmange, G.; Slavov, N. SCoPE-MS: Mass Spectrometry of Single Mammalian Cells Quantifies Proteome Heterogeneity During Cell Differentiation. *Genome Biol.* **2018**, *19* (1), 161.

93. Specht, H.; Emmott, E.; Petelski, A. A.; Huffman, R. G.; Perlman, D. H.; Serra, M.; Kharchenko, P.; Koller, A.; Slavov, N. Single-Cell Proteomic and Transcriptomic Analysis of Macrophage Heterogeneity Using SCoPE2. *Genome Biol.* **2021**, *22* (1), 50.
94. Yang, M.; Nelson, R.; Ros, A. Toward Analysis of Proteins in Single Cells: A Quantitative Approach Employing Isobaric Tags with MALDI Mass Spectrometry Realized with a Microfluidic Platform. *Anal. Chem.* **2016**, *88* (13), 6672–6679.
95. Roy Swapan, K.; Kim, S.; Yoon Jung, H.; Yoon, Y.-K.; Cho, K. Recent Progress on Microfluidic Electrophoresis Device Application in Mass Spectrometry. *Mass Spectrom. Lett.* **2018**, *9* (1), 1–16.
96. Xue, Q. F.; Foret, F.; Dunayevskiy, Y. M.; Zavracky, P. M.; McGruer, N. E.; Karger, B. L. Multichannel Microchip Electrospray Mass Spectrometry. *Anal. Chem.* **1997**, *69* (3), 426–430.
97. Ramsey, R. S.; Ramsey, J. M. Generating Electrospray from Microchip Devices Using Electroosmotic Pumping. *Anal. Chem.* **1997**, *69* (13), 2617.
98. Zhu, Y.; Piehowski, P. D.; Zhao, R.; Chen, J.; Shen, Y.; Moore, R. J.; Shukla, A. K.; Petyuk, V. A.; Campbell-Thompson, M.; Mathews, C. E.; Smith, R. D.; Qian, W. J.; Kelly, R. T. Nanodroplet Processing Platform for Deep and Quantitative Proteome Profiling of 10-100 Mammalian Cells. *Nat. Commun.* **2018**, *9* (1), 882.
99. Mellors, J. S.; Jorabchi, K.; Smith, L. M.; Ramsey, J. M. Integrated Microfluidic Device for Automated Single Cell Analysis Using Electrophoretic Separation and Electrospray Ionization Mass Spectrometry. *Anal. Chem.* **2010**, *82* (3), 967–973.
100. Yang, M.; Chao, T.-C.; Nelson, R.; Ros, A. Direct Detection of Peptides and Proteins on a Microfluidic Platform with MALDI Mass Spectrometry. *Anal. Bioanal. Chem.* **2012**, *404* (6), 1681–1689.
101. Li, Z. Y.; Huang, M.; Wang, X. K.; Zhu, Y.; Li, J. S.; Wong, C. C. L.; Fang, Q. Nanoliter-Scale Oil-Air-Droplet Chip-Based Single Cell Proteomic Analysis. *Anal. Chem.* **2018**, *90* (8), 5430–5438.
102. Zhu, Y.; Podolak, J.; Zhao, R.; Shukla, A. K.; Moore, R. J.; Thomas, G. V.; Kelly, R. T. Proteome Profiling of 1 to 5 Spiked Circulating Tumor Cells Isolated from Whole Blood Using Immunodensity Enrichment, Laser Capture Microdissection, Nanodroplet Sample Processing, and Ultrasensitive nanoLC-MS. *Anal. Chem.* **2018**, *90* (20), 11756–11759.
103. Zhu, Y.; Clair, G.; Chrisler, W. B.; Shen, Y.; Zhao, R.; Shukla, A. K.; Moore, R. J.; Misra, R. S.; Pryhuber, G. S.; Smith, R. D.; Ansong, C.; Kelly, R. T. Proteomic Analysis of Single Mammalian Cells Enabled by Microfluidic Nanodroplet Sample Preparation and Ultrasensitive NanoLC-MS. *Angew. Chem. Int. Ed. Engl.* **2018**, *57* (38), 12370–12374.
104. Dou, M.; Clair, G.; Tsai, C. F.; Xu, K.; Chrisler, W. B.; Sontag, R. L.; Zhao, R.; Moore, R. J.; Liu, T.; Pasa-Tolic, L.; Smith, R. D.; Shi, T.; Adkins, J. N.; Qian, W. J.; Kelly, R. T.; Ansong, C.; Zhu, Y. High-Throughput Single Cell Proteomics Enabled by Multiplex Isobaric Labeling in a Nanodroplet Sample Preparation Platform. *Anal. Chem.* **2019**, *91* (20), 13119–13127.
105. Darmanis, S.; Gallant, C. J.; Marinescu, V. D.; Niklasson, M.; Segerman, A.; Flamourakis, G.; Fredriksson, S.; Assarsson, E.; Lundberg, M.; Nelander, S.; Westermark, B.; Landegren, U. Simultaneous Multiplexed Measurement of RNA and Proteins in Single Cells. *Cell Rep.* **2016**, *14* (2), 380–389.
106. DeLaughter, D. M. The Use of the Fluidigm C1 for RNA Expression Analyses of Single Cells. *Curr. Protoc. Mol. Biol.* **2018**, *122* (1), e55.
107. Genshaft, A. S.; Li, S.; Gallant, C. J.; Darmanis, S.; Prakadan, S. M.; Ziegler, C. G.; Lundberg, M.; Fredriksson, S.; Hong, J.; Regev, A.; Livak, K. J.; Landegren, U.; Shalek, A. K. Multiplexed, Targeted Profiling of Single-Cell Proteomes and Transcriptomes in a Single Reaction. *Genome Biol.* **2016**, *17* (1), 188.

108. Macaulay, I. C.; Ponting, C. P.; Voet, T. Single-Cell Multiomics: Multiple Measurements from Single Cells. *Trends Genet.* **2017**, *33* (2), 155–168.
109. Marx, V. A Dream of Single-Cell Proteomics. *Nat. Methods* **2019**, *16* (9), 809–812.
110. Wu, Z.; Lawrence, P. J.; Ma, A.; Zhu, J.; Xu, D.; Ma, Q. Single-Cell Techniques and Deep Learning in Predicting Drug Response. *Trends Pharmacol. Sci.* **2020**, *41* (12), 1050–1065.

CHAPTER 3

Biosensors for single-cell metabolomic characterization

Xingxiu Chen and Zhibo Yang*
Department of Chemistry and Biochemistry, Stephenson Life Sciences Research Center, University of Oklahoma, Norman, OK, United States

1 Introduction

Cell is the most fundamental unit of all living organisms. Heterogeneous cells with distinct morphology features and phenotypic profiles coexist in biological systems, implying the collective study of a population of cells cannot represent the behavior of individual cells. The importance of cell heterogeneity has been appreciated in many studies of diseases, such as cancers, infectious diseases, and brain disorders, as human body is composed of various cell types with unique functions and in different environments.[1] Studies of cellular heterogeneity have been performed in different areas, including genomics, transcriptomics, proteomics, and metabolomics. Metabolites, the downstream products of cell metabolism, are small biomolecules (molecular weight <2 kDa) directly involved in numerous cellular functions such as metabolism, growth, and reproduction of cells. Biological heterogeneity of cells can be revealed by studying their metabolites at the single-cell level. The major challenges in single-cell metabolomics studies include rapid turnover rate, low concentrations, limited volume of sample, and high structural diversity.[2,3] Fortunately, recent development of highly sensitive analytical techniques enables us to perform single-cell metabolomics analyses. A variety of analytical techniques, including florescence, NMR (nuclear magnetic resonance), electrochemical detection, microfluidic device, and mass spectrometry (MS), have been used for the research of single-cell metabolomics. This chapter focuses on reviewing recent development of techniques, with an emphasis on MS-based methods, for single-cell metabolites studies.

2 Fluorescence

Fluorescence has been widely used in a variety of sensitive analytical techniques to visualize, track, and quantify both large cellular proteins and small

*Corresponding author.

https://doi.org/10.1016/B978-0-323-89841-6.00001-3

signaling molecules in cell analysis such as live-cell imaging and single-molecule detection. The early application of fluorescence-based single cell study is mostly focus on monitoring proteins (e.g., profiling proteins by fluorescence flow cytometry and protein imaging by fluorescence single-cell microscopy), whereas its capabilities to measure different cellular metabolites are reported in recent publications.[4–6]

The prerequisite of performing fluorescence-based detection of metabolites in single cells is the presence of fluorescent moieties, which are either from endogenous fluorescence metabolites or exogenous molecules with fluorescent tag.[7] As a label-free, nondestructive analysis technique, autofluorescence detection is appealing in single-cell analysis. However, there are only limited numbers of autofluorescent metabolites, including reduced nicotinamide adenine dinucleotide (NADH),[8–11] flavin mononucleotide (FMN) and flavin adenine dinucleotide (FAD)[11, 12], carotenoids,[13] chlorophyll pigment,[14] and drug compounds (e.g., quinolone[15] and doxorubicin[16]), can be directly studied in single cells. In addition, the fluorescent readout can be interfered by a wide range of factors in the microenvironment, such as the pH of cell growth medium and cell density, potentially impacting the accuracy of detection and further complicating the interpretation.[17]

To enable the autofluorescence measurement in single cells, miniaturized sampling techniques have been developed. For example, antibody-based fiber-optic nanoprobe has been fabricated to insert into individual cells for measuring fluorescent carcinogen compound.[18, 19]

Most fluorescence-based single-cell analyses require the attachment of fluorescent tag during sample preparation, and common methods include fluorescent protein fusion and chemical labeling of fluorescent moiety. Typically, fluorescent protein fusion requires the production of genetically encoded sensors, which normally comprise one sensor protein (e.g., bacterial protein) binding to specific metabolites and one (or two) fluorescent proteins.[20] Binding of the target metabolite on the sensor protein can result in changed structures in fluorescent protein and altered fluorescent signals. Currently, genetically encoded sensors have been used to monitor a few metabolites (e.g., ATP,[21, 22] glucose,[23] lactate,[24] guanosine 3′,5′-cylclic monophosphate[25]) in single cells. In addition, Förster resonance energy transfer (FRET), in which two fluorescent proteins (fluorescent donor and acceptor) are used in genetically encoded sensors, can be employed to measure the fluorescent ratio change to determine the ligand concentration. Specifically, metabolites binding to the sensor proteins can lead to changes in the distance between two fluorophores in fluorescent proteins, and then alter the FRET efficiency. Another approach to tagging target metabolites with fluorescence is using chemical reactions to attach fluorescent moiety, and a large number of fluorescent dye-based probes have been designed and synthesized.[26] However, this approach to measuring intracellular metabolites is restricted by a few potential limitations such as the biological toxicity and membrane permeability of probes as well as affected functions of metabolites due to structure modification.[27]

The applications of fluorescence-based methods are primarily restricted by the limited number of metabolites that can be simultaneously detected in single cells. To overcome this limit, multiple fluorescent probes have been used.[28] However, multiple-staining processes require the conjugation of several fluorescent probes, which may result in fluorescent quenching due to molecular interactions among them.[29] In addition, fluorescent biosensors have been utilized in other analytical techniques, such as microfluidic device, to extend the capabilities of detecting more molecules. For example, the single-cell barcode chip (SCBC), an integrated device to quantity metabolites and proteins in single cells, incorporates the surface-competitive binding assay into microfluidic chip to continuously achieve single-cell separation, on-chip cell lysis, metabolite competition assays, and protein immunofluorescence assay (Fig. 1A).[30] Using this platform, a number of metabolites and proteins were captured by their corresponding surface-immobilized antibodies, followed by quantification through fluorescence measurements.[30] In another study, a microfluidic device combining microchip electrophoresis and laser-induced fluorescence detection has been developed. This multicolor fluorescence detection method has been used to study multiple metabolites in single cells (Fig. 1B).[31] Target metabolites are labeled by multiple fluorescent probes in single cells. Then, using this integrated device, single-cell separation, cell lysis, metabolites separation, and detection of multicolor fluorescently labeled metabolites can be achieved.[31]

3 Electrochemical detection

Electrochemical analysis is based on the measurements of electrochemically active molecules. This method has been used as an important tool to detect neurotransmitters, such as norepinephrine, epinephrine, and dopamine, in the studies of neurotransmitter dynamics.[32] The development of miniaturized electrodes (e.g., ultramicroelectrode and nanoelectrode) enables both qualitative and quantitative analysis at the single-cell level. Common electrochemical techniques used in single-cell metabolomics study, including

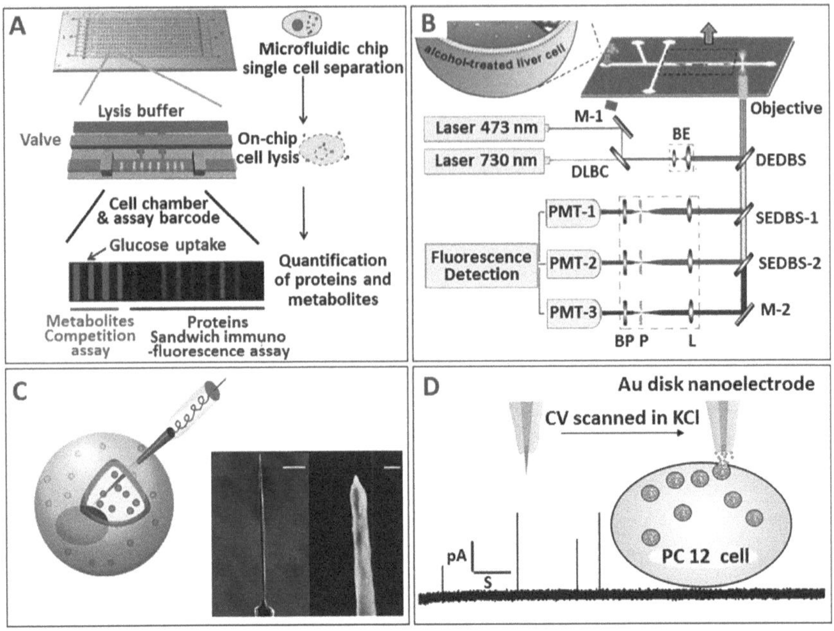

Fig. 1 Examples of fluorescence- and electrochemistry-based techniques for single-cell metabolomics studies. (A) Schematic drawing of single-cell barcode chip capable of integrated quantitative analyses of proteins and metabolites. (B) Instrumental setup of the multicolor fluorescence detection-based microfluidic system. (C) Nanotip canonical carbon-fiber microelectrode used for the detection of transmitters in lysed cytoplasm vesicles from single PC12 cells. (D) An Au disk nanoelectrode with a shape tip (radius down to 3 nm) used to monitor dopamine release from single PC12 cells. *((Panel A) Reproduced from reference Xue, M.; Wei, W.; Su, Y. P.; Kirn, J.; Shin, Y. S.; Mai, W. X.; Nathanson, D. A.; Heath, J. R. Chemical Methods for the Simultaneous Quantitation of Metabolites and Proteins from Single Cells. J. Am. Chem. Soc.* ***2015****, 137 (12), 4066–4069, with permission from American Chemical Society. (Panel B) Reproduced from reference Li, Q. L.; Chen, P. L.; Fan, Y. Y.; Wang, X.; Xu, K. H.; Li, L.; Tang, B. Multicolor Fluorescence Detection-Based Microfluidic Device for Single-Cell Metabolomics: Simultaneous Quantitation of Multiple Small Molecules in Primary Liver Cells. Anal. Chem.* ***2016****, 88 (17), 8610–8616, with permission from American Chemical Society. (Panel C) Reproduced from reference Li, X. C.; Majdi, S.; Dunevall, J.; Fathali, H.; Ewing, A. G. Quantitative Measurement of Transmitters in Individual Vesicles in the Cytoplasm of Single Cells with Nanotip Electrodes. Angew. Chem. Int. Ed.* ***2015****, 54 (41), 11978–11982, with permission from John Wiley and Sons. (Panel D) Reproduced from reference Liu, Y. Z.; Li, M. N.; Zhang, F.; Zhu, A. W.; Shi, G. Y. Development of Au Disk Nanoelectrode Down to 3 nm in Radius for Detection of Dopamine Release from a Single Cell. Anal. Chem.* ***2015****, 87 (11), 5531–5538, with permission from American Chemical Society.)*

amperometry and fast-scan cyclic voltammetry (FSCV), rely on redox reactions on the electrode surfaces. Versatile materials have been utilized to fabricate substrates (e.g., using silicon, glass, and polymers) and electrodes (e.g., using carbon, carbon fiber, carbon nanotubes, and metals), as detailed in published review articles.[33, 34] Li et al. developed flame-etched carbon-fiber microelectrodes (tip diameter: 50–100 nm) to quantify catecholamines in individual, nanoscale vesicles of single PC12 cells (Fig. 1C).[35] Shi and coworker designed Au disk nanoelectrode (with 3 nm as the smallest radius), which showed comparable amperometric sensitivity as carbon fiber electrode, to monitor dopamine released from single vesicles of single cells (Fig. 1D).[36] In addition to measuring extracellular metabolites' concentrations (with the electrode positioned at the outer surface of single cell to detect released neurotransmitters during exocytotic events), ultrasmall electrodes also allow for monitoring intracellular metabolites in real-time (with the electrode inserted into single cell). The tip size of microelectrodes is crucial for the measurement of intracellular metabolites, as small tips can minimize cell damage when being inserted into the cell.

To enable electrochemical analysis of nonelectroactive species at the single-cell level, researchers functionalized electrode surface with enzyme to target the analytes of interest. The fabrication of enzyme-modified electrodes requires either covalent attachment of the enzyme to electrode surface or film coating. Using enzyme specifically reacting with target analytes, hydrogen peroxides can be generated as electroactive reporter molecules for the electrochemical measurement. A variety of molecules, including glucose,[37] glutamate,[38] NO,[39] and glutathione,[40] have been detected as nonelectroactive compounds at the single-cell level. As a highly sensitive technique, electrochemical measurement has also been coupled to other single-cell analysis methods. For example, amperometric detection has been used for metabolites quantification in microfluidic chips that integrate cell injection, cell lysis, and metabolites detection for single-cell analysis.[41] To test the devices, ascorbic acid was selected as a model compound for concentration measurement in single wheat callus cells. Pan et al. designed the nanokit by adding the kit components (i.e., sphingomyelin, glucose oxidase, choline oxidase, and alkaline phosphatese) into a nanometer-sized capillary with a ring electrode at the tip.[42] Authors injected the nanokit into single cells, pumped these compounds out, and then used coulometry to measure the intracellular glucose concentrations and sphingomyelinase activities.

4 Mass spectrometry

Mass spectrometry (MS), owing to its high sensitivity, multiplexing capabilities, high analyte coverages, and excellent versatility, has become the major analytical platform for both targeted and untargeted metabolomics studies, including those at the single-cell level. The development of instrumentation, such as sampling device, ionization method, and mass analyzer, plays a key role in single-cell metabolomics study.

Recent advancement in sampling and ionization methods enables MS to study various types of cells, ranging from small bacteria (e.g., *Escherichia coli*, cell length < 2.0 μm) to large embryos (e.g., *Xenopus laevis* embryo, cell diameter ~1 mm) under different conditions.[43–45] Based on the sampling and ionization environment, single-cell MS (SCMS) metabolomics experiments can be performed under vacuum condition, in which cells need to be dehydrated before the measurement, or in ambient environment, in which live cells can be directly analyzed in their native or near-native status. Although many types of mass analyzers can be utilized in single-cell MS platforms, their selection is generally determined by the purposes of studies. The commonly used mass analyzers in the reported studies include the TOF (time of flight), ion mobility spectrometer, Fourier-transform ion cyclotron resonance (FT-ICR), and Orbitrap. These mass analyzers provide different advantages: the TOF offers fast scan rate and broad mass range, ion mobility spectrometer enables the separation of metabolites (e.g., isomers) in the gas-phase, and the FT-ICR as well as Orbitrap have high mass accuracy and high mass resolution.

4.1 Vacuum-based single-cell mass spectrometry methods

Secondary ion MS (SIMS) and matrix-assisted laser desorption ionization (MALDI) are common approaches to conducting single-cell metabolomics experiments under vacuum sampling and ionization conditions. These methods have been commercialized, and they provide high spatial resolution and excellent sensitivities for single-cell measurement.

4.1.1 Secondary ion mass spectrometry

The operation of SIMS relies on the detection of secondary ions sputtered by bombarding the sample surface with a focused, highly energetic primary ion beam. The major advantages of SIMS include low detection limit, high spatial resolution, and capability of depth analysis. This technique has been utilized for metabolomics analysis at cellular and subcellular levels, allowing for single-cell imaging.[46–48] However, the major disadvantage of SIMS is that highly fragmented ions are produced, and data analysis of large numbers of biomolecules is challenging. To reduce ion fragmentation while increasing the sensitivity, devices producing ion clusters and polyatomic primary ion beams, including argon gas cluster ion gun, bismuth cluster ion gun, and C_{60}^{+} primary ion source, have been widely used in single-cell metabolomics studies. In addition, sample preparation methods (e.g., surface modification) have been developed to minimize ion fragmentation. Previous studies indicate that ionic liquid matrix-enhanced SIMS can enhance the signals of certain intact lipids (Fig. 2A),[49] and metal-assisted SIMS can enhance the signal intensities of intact cholesterol and lipid molecules in single-cell studies.[50]

There are a variety of different applications of using TOF-SIMS to qualitatively and quantitatively study metabolites, lipids, and drugs in single cells. This method has been utilized to classify bacterial phenotypes[51] and to distinguish different breast cancer cell lines based on their differences of chemical profiles.[52] Owing to its high spatial resolution (e.g., below 100 nm[52]), SIMS is well-suited for MS imaging studies of single cells. In particular, the state of art systems of nanoscale secondary ion mass spectrometry (nano-SIMS) has achieved a spatial resolution down to 50 nm. Combined with its capability of depth analysis, SIMS enables mapping the 3D distribution of metabolites, small molecules, lipids, and drug compounds in single cells. For example, Ostrowski et al. used SIMS imaging to determine cholesterol concentration in plasma membranes of single macrophage cells, and the obtained SIMS images were correlated with fluorescence images to confirm chemical differences between drug-treated and control J774 cells (Fig. 2B).[53] In another study, 3D SIMS imaging experiment was combined with multivariate data analysis to discriminate different cell compartments, including cell nucleus, intracellular space, and cell membrane, inside single HeLa cells.[54] Quantitative studies of different cell compartments can be achieved using cells labeled by stable isotopes. Jiang et al. combined nano-SIMS imaging with backscattered electron imaging to quantify ^{15}N-labeled glutamine in subcellular structures such as nucleus, endoplasmic reticulum, and Golgi apparatus.[55]

4.1.2 Matrix-assisted laser desorption ionization

MALDI uses intense laser pulses to ablate the sample incorporated with matrix molecules for desorption and ionization of analytes. This technique has been widely used in MS imaging studies of proteins, owing to

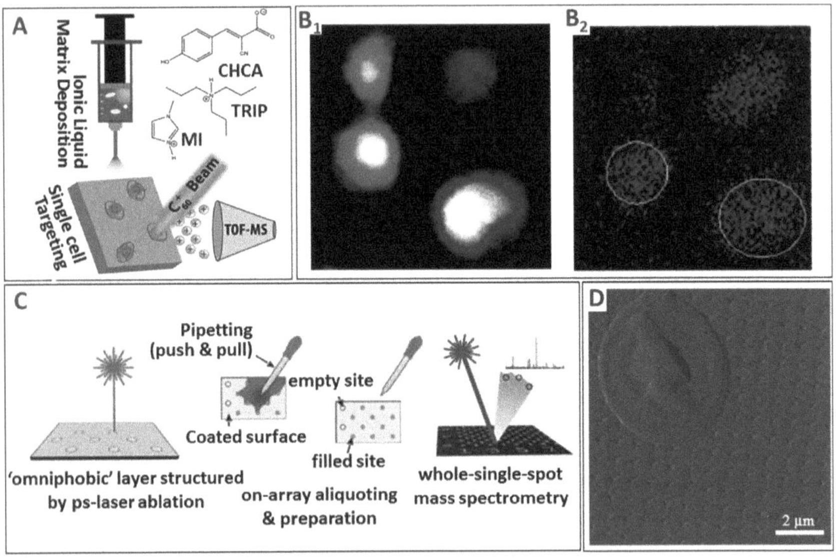

Fig. 2 Examples of vacuum-based MS techniques for single-cell metabolomics studies. (A) Using ionic liquid for matrix-enhanced SIMS analysis of lipids.[49] (B) SIMS studies. (B_1) Control (*green—light grey in print version*) and cholesterol-treated (*red—grey in print version*) J774 cells, which are illustrated by two-color fluorescence images, were analyzed by SIMS showing the images of $C_5H_9^+$ from single cells. (C) Experimental workflow of using microarrays for high-throughput single cell MALDI-MS analysis. (D) Atomic force microscope (AFM) image of a single yeast cell on the NAPA prepared for matrix-free desorption ionization MS analysis. *((Panel A) Reproduced from reference Do, T. D.; Comi, T. J.; Dunham, S. J. B.; Rubakhin, S. S.; Sweedler, J. V. Single Cell Profiling Using Ionic Liquid Matrix-Enhanced Secondary Ion Mass Spectrometry for Neuronal Cell Type Differentiation. Anal. Chem.* ***2017****, 89 (5), 3078–3086, with permission from American Chemical Society. (Panel B) Reproduced from reference Ostrowski, S. G.; Kurczy, M. E.; Roddy, T. P.; Winograd, N.; Ewing, A. G. Secondary Ion MS Imaging to Relatively Quantify Cholesterol in the Membranes of Individual Cells from Differentially Treated Populations. Anal. Chem.* ***2007****, 79 (10), 3554–3560, with permission from American Chemical Society. (Panel C) Reproduced from reference Urban, P. L.; Jefimovs, K.; Amantonico, A.; Fagerer, S. R.; Schmid, T.; Madler, S.; Puigmarti-Luis, J.; Goedecke, N.; Zenobi, R. High-Density Micro-Arrays for Mass Spectrometry. Lab Chip* ***2010****, 10 (23), 3206–3209, with permission from Royal Society of Chemistry. (Panel D) Reproduced from reference Walker, B. N.; Stolee, J. A.; Vertes, A. Nanophotonic Ionization for Ultratrace and Single-Cell Analysis by Mass Spectrometry. Anal. Chem.* ***2012****, 84 (18), 7756–7762, permission from American Chemical Society.)*

its wide mass range coverage, and metabolites. Recent developments in the application of MALDI MS in single-cell metabolomics study include applying novel sampling devices to increase the analytical throughput of profiling metabolites in cells and optimizing sample preparation condition to decrease the influence of matrix. Combined with other techniques,

MALDI allows for high-throughput analyses of single cell. Ong et al. used an optical microscopy to record cell coordinates of single cells, followed by automated MALDI-TOF analysis. This method significantly reduced data acquisition time when analyzing a large number of cells (e.g., 3000 cells were analyzed within 1 h) compared with typical MALDI MS (e.g., several hundred hours).[56] In addition, microarray substrates, which are composed of hydrophilic reservoirs on an omniphobic surface, have been designed to automatedly aliquot cell suspension by trapping uniform droplets (contain single cells) into the hydrophilic reservoirs (Fig. 2C), facilitating high-throughput MALDI-MS analysis (i.e., more than two samples per second).[57, 58]

Although matrixes are needed for MALDI experiments, they generally produce significant background noises at low-mass range, complicating the spectrum interpretation of metabolites. To circumvent this limitation, several techniques have been developed to conduct matrix-free desorption ionization MS studies. Desorption ionization on silicon (DIOS) uses porous silicon as the substrate to absorb UV laser for the desorption and ionization of analytes. Kruse et al. reported protocols to directly culture invertebrate neurons on the surface of porous silicon, and they detected peptides from individual neurons using DIOS-MS analysis.[59] Although the development of nanostructure-initiator MS (NIMS) was based on DIOS-MS, initiator compounds were used to promote desorption and ionization of analytes.[60] Specifically, NIMS substrate contains nanoporous silicon, in which initiators are trapped. Upon being irradiated by a UV laser, rapid heating of porous silicon surface causes vaporization of the initiator, which further improves the desorption/ionization of analytes. Because of the absence of traditional matrix, NIMS facilitates the analysis of metabolites in low-mass range. For example, O'Brien et al. successfully applied NIMS to detect chemotherapy drugs and endogenous metabolites in single cells.[61] Nanopost array structure (NAPA) is another example of matrix-free substrate in MS studies. NAPA has a structure of columnar silicon nanopost array with precisely produced height, diameter, and periodicity. In particular, the optimized aspect ratio (height/diameter) and subwavelength post diameters of NAPA increased ion yields of small molecules via the resonance-like enhancement (Fig. 2D).[62] For example, 24 biologically significant metabolites, accounting for 4% of the yeast metabolome, were detected in a single yeast cell using NAPA laser desorption ionization MS method.[62]

4.2 Ambient single-cell mass spectrometry methods

Metabolomics studies at the single-cell level can be carried out in ambient conditions (i.e., room temperature and atmospheric pressure) with no or minimum sample preparation, allowing for cells to be studied in their native or near-native environment. Recent development of ambient sampling and ionization techniques greatly fostered direct, real-time analysis of living single cells, and improved our understanding of the functions and regulations of cellular metabolism.[63, 64] Numerous ambient sampling and ionization techniques have been developed for single-cell metabolomics studies. Here, we summarized these methods into three major categories: probe-based, laser-based, and microfluidics-based techniques. In addition, separation methods (either before or after ionization) and spectroscopy techniques can be coupled to ambient SCMS techniques to acquire additional advantages (e.g., increased sensitivity and extra structure information).

4.2.1 Probe-based techniques

The structures of the sampling and ionization probes are the cores of probe-based techniques. Depending the channel numbers inside these probes, we tentatively classify them into three major categories: solid probes (without channel), monochannel probes (with one channel), and multichannel probes (with two or more channels). The general working mechanisms of probes in these three categories are similar: cellular contents are extracted by a probe followed by nano-ESI-based ionization and MS detection.

Solid probes

These probes are designed as solid metal needles without inner channel, and experiments are generally performed in two consecutive steps: cellular contents sampling and MS analysis. Direct microsampling is generally achieved by extracting cellular contents onto the surface of a solid metal needle with the tip size smaller than a cell. Hiraoka and coworkers developed probe electrospray ionization (PESI), which employs a solid needle as both the microsampling probe and the electrospray emitter.[65] Surface modification of PESI probes has been carried out to improve their performance. Solid phase microextraction (SPME) has been used as a separation method to extract analytes by a modified solid support, and the absorbed analytes can be desorbed (e.g., using thermal heating and solvent) for analysis.[66] PESI probes prepared using metal needles with modified surface have been utilized to enrich metabolites in single cells. Gong et al. produced tungsten probes to conduct qualitative and quantitative metabolites in subcellular compartment of *Allium cepa* cells (Fig. 3A).[67] To decease the solvent volume on the probe tip and subsequently increase the sensitivity, Yu et al. prepared hydrophilic surface (via oxidative treatment) on stainless-steel sampling probes, and they used a piezoelectric inkjet system to provide auxiliary solvent and assist ESI process (Fig. 3B).[68] Chen et al. developed a single-cell MS screening platform by combing drop-on-demand inkjet cell printing with PESI MS (Fig. 3C).[69] By aligning the inkjet nozzle with the tungsten tip, droplets containing single cells are sprayed onto the needle and subsequently ionized. Recently, Zheng et al. functionalized copper probes using reduced graphene oxide to increase their surface area for improved enrichment of analytes. In addition, a nitrogen aggregation/gas heating system was introduced to promote desorption of absorbed metabolites and facilitate ion transport for improved detection sensitivity.[70]

Monochannel probes

A monochannel probe contains an inner channel, and this type of device is used as both sampling needle and nano-ESI emitter to directly analyze extracted cellular contents from selected single cells. Accordingly, experiments are generally performed in two consecutive steps: cellular contents sampling and MS analysis. Direct micro-sampling is generally achieved by drawing cellular contents through micropipettes with sharp tips smaller than cell size. Collected cellular contents will then be ionized though nano-ESI for MS analysis. These probes are generally produced using quartz/fused silica and borosilicate glass. Typical examples of these techniques include PESI MS,[67, 71] live single-cell MS,[72] capillary micropipette with induced ESI MS,[73, 74] and micropipette needle[75]. Although the PESI probes were originally produced as solid metal needles, probes containing an inner channel have been fabricated using nonmetal materials such as quartz and glass.[71, 76] In fact, these types of PESI probes are essentially nano-ESI emitters.

Live single-cell MS utilizes a metal-coated glass nanospray emitter, which is used as both the micro-sampling device and nano-ESI source (Fig. 4A).[77, 78] Cellular contents are sucked into the nanospray emitter using a micromanipulator, and then a small amount of organic solvent is added and mixed with cellular contents. The nano-ESI emitter is then connected to an ionization source for MS analysis. Applications of Live single-cell MS include detecting endogenous metabolites, extracellular compounds, and drug and drug metabolites from single cells and subcellular compartments.[79] To analyze nonadherent cells, sonication can be used to improve lysis efficiency and assist mixing of lysate and organic solvent, resulting in increased detection

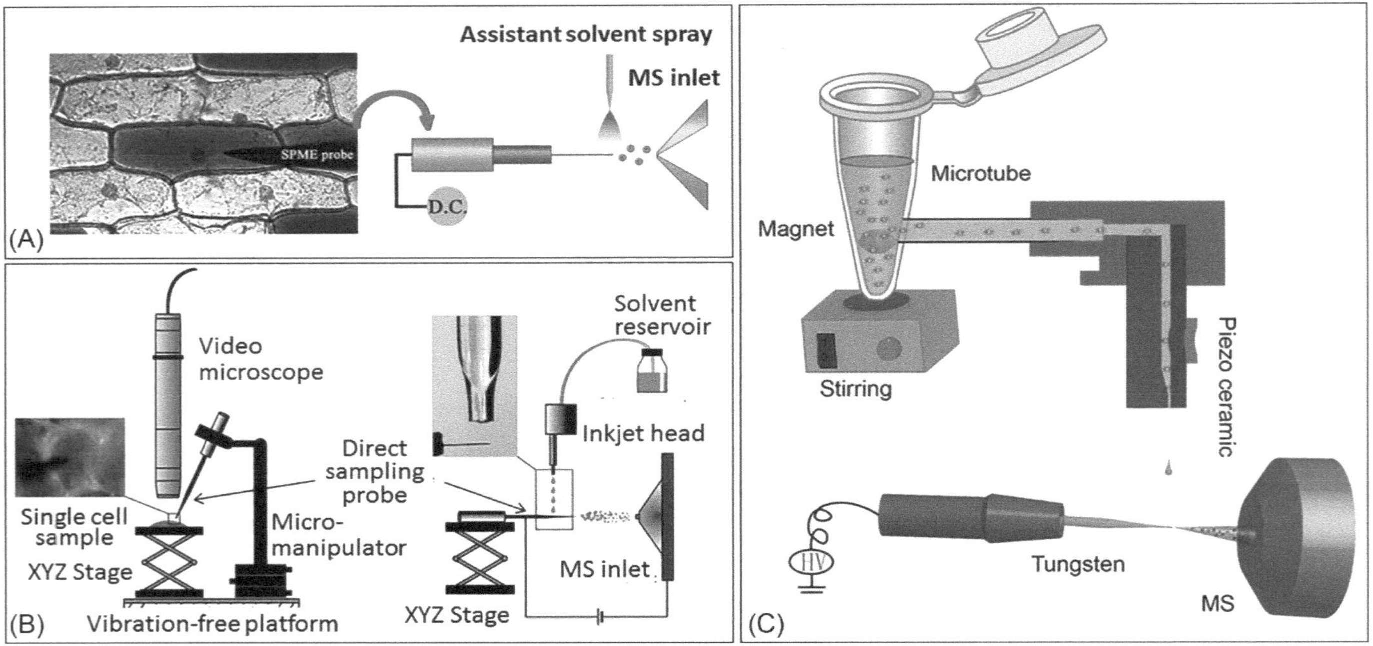

Fig. 3 Examples of solid probes used in the probe-based SCMS metabolomics techniques. (A) Using PESI-MS to analyze a single *Allium cepa* cells. The tungsten-made probe was used as sampling probe and ESI emitter. (B) The instrumental setup for direct sampling probe MS. A stainless steel with hydrophilic surface was used for sampling, and ionization of metabolites extracted on needle surface was assisted by solvent ejected from a piezoelectric inkjet system. (C) Integrated inkjet cell printing and PESI-MS setup. Droplets containing single cells were generated from the inkjet nozzle cell manipulator and dripped on the tungsten probe for ionization and MS analysis. *((Panel A) Reproduced from reference Gong, X. Y.; Zhao, Y. Y.; Cai, S. Q.; Fu, S. J.; Yang, C. D.; Zhang, S. C.; Zhang, X. R. Single Cell Analysis with Probe ESI-Mass Spectrometry: Detection of Metabolites at Cellular and Subcellular Levels. Anal. Chem.* ***2014****, 86 (8), 3809–3816, with permission from American Chemical Society. (Panel B) Reproduced from reference Yu, Z.; Chen, L. C.; Ninomiya, S.; Mandal, M. K.; Hiraoka, K.; Nonami, H. Piezoelectric Inkjet Assisted Rapid Electrospray Ionization Mass Spectrometric Analysis of Metabolites in Plant Single Cells Via a Direct Sampling Probe. Analyst* ***2014****, 139 (22), 5734–5739, with permission from Royal Society of Chemistry. (Panel C) Reproduced from reference Chen, F. M.; Lin, L. Y.; Zhang, J.; He, Z. Y.; Uchiyama, K.; Lin, J. M. Single-Cell Analysis Using Drop-on-Demand Inkjet Printing and Probe Electrospray Ionization Mass Spectrometry. Anal. Chem.* ***2016****, 88 (8), 4354–4360, with permission from American Chemical Society.)*

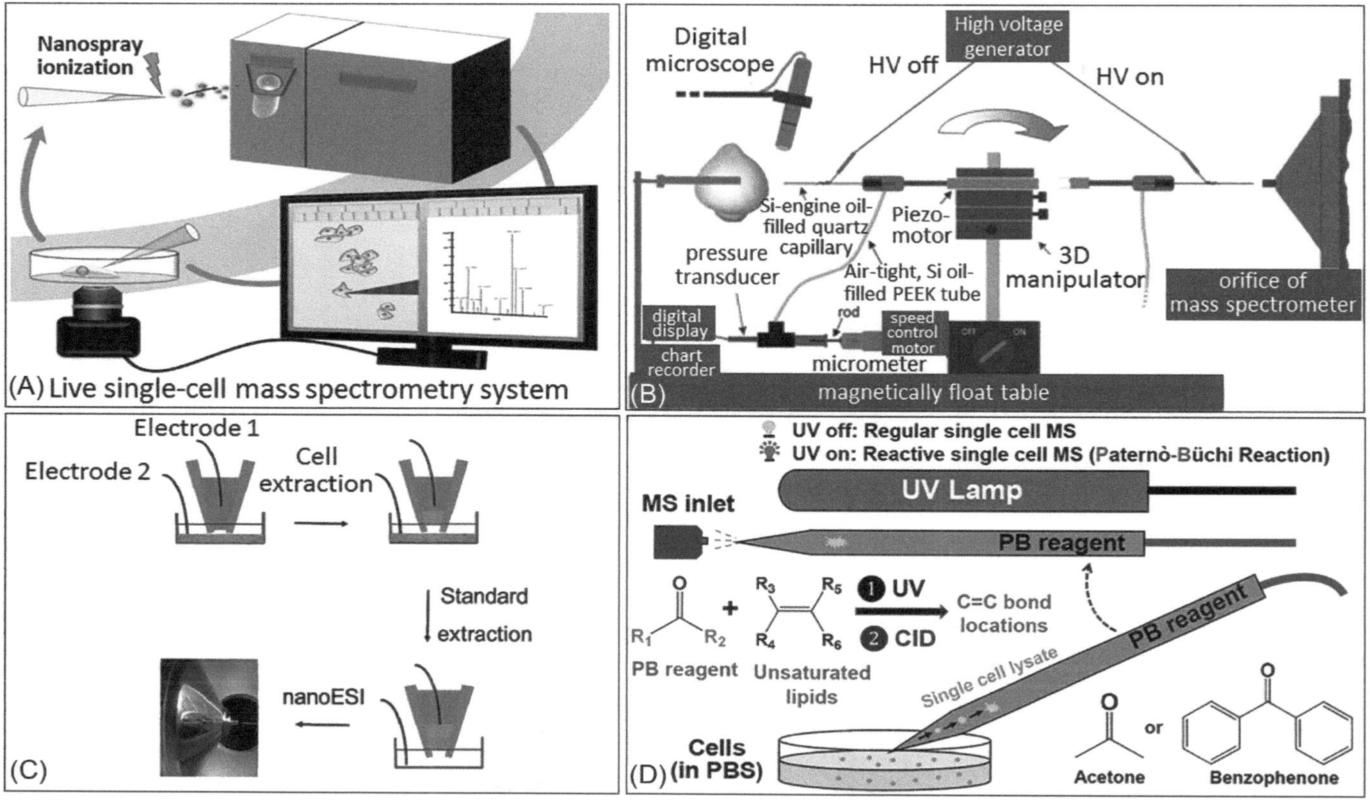

See the legend on next page.

Fig. 4 Examples of monochannel probes used in the probe-based SCMS metabolomics techniques. (A) The experiment design of Live single-cell MS. A metal-coated nano-ESI emitter was used to suck cellular contents controlled by micromanipulator, and the emitter was then transferred to a mass spectrometer for analysis. (B) The instrumental design of the cell pressure probe. Single-cell sampling was achieved by inserting a quartz probe into single cell prior to MS detection. (C) Combining electroosmotic extraction with MS for single-cell metabolomics study. Electrode 1 was inserted into a nanopipette containing electrolyte solution, and electrode 2 was placed onto the sample. By applying voltage between these two electrodes, cell contents were extracted into the nanopipette for subsequent MS analysis. (D) Using micropipette needles to determine C=C bond positions in unsaturated lipids in single cells. The glass micropipette needle was utilized as the single-cell sampling probe, PB reaction container, and nano-ESI emitter. *((Panel A) Reproduced from reference Abouleila, Y.; Onidani, K.; Ali, A.; Shoji, H.; Kawai, T.; Lim, C. T.; Kumar, V.; Okaya, S.; Kato, K.; Hiyama, E.; Yanagida, T.; Masujima, T.; Shimizu, Y.; Honda, K. Live Single Cell Mass Spectrometry Reveals Cancer-Specific Metabolic Profiles of Circulating Tumor Cells. Cancer Sci.* ***2019****, 110 (2), 697–706, with permission from John Wiley and Sons. (Panel B) Reproduced from reference Gholipour, Y.; Erra-Balsells, R.; Hiraoka, K.; Nonami, H. Living Cell Manipulation, Manageable Sampling, and Shotgun Picoliter Electrospray Mass Spectrometry for Profiling Metabolites. Anal. Biochem.* ***2013****, 433 (1), 70–78, with permission from Elsevier. (Panel C) Reproduced from reference Yin, R. C.; Prabhakaran, V.; Laskin, J. Quantitative Extraction and Mass Spectrometry Analysis at a Single-Cell Level. Anal. Chem.* ***2018****, 90 (13), 7937–7945, with permission from American Chemical Society. (Panel D) Reproduced from reference Zhu, Y. L.; Wang, W. H.; Yang, Z. B., Combining Mass Spectrometry with Paterno-Buchi Reaction to Determine Double-Bond Positions in Lipids at the Single-Cell Level. Anal. Chem.* ***2020****, 92 (16), 11380–11387, with permission from American Chemical Society.)*

of metabolites.[80] Live single-cell MS technique has been combined with other imaging tools, such as fluorescence imaging, holographic, and tomographic laser microscopy, to target organelles for metabolomics analysis and to perform quantitative studies of single cells.[81, 82] Nakashima et al. used quartz capillaries to produce PESI probes, which have similar designs as the nano-ESI emitters. They added an ionic liquid and placed a titanium wire electrode inside the probe to improve electric conductivity and increase MS detection sensitivity.[76] Zhao et al. coupled a glass pressure-assisted microsampling probe with hydrogen flame desorption ionization (HFDI) to achieve direct quantitation of metabolites in single cells.[83]

Induced nano-electrospray ionization (induced nESI), developed by Cooks and coworkers, is an efficient ionization technique with a high tolerance to matrix effect. In this design, electrospray of analytes is induced by placing an electrode, which carries pulsed potential, close to the emitter (e.g., <2 mm).[84] The induced nESI ionization method has been modified to improve its performance in single-cell analysis. To enhance the detection sensitivity of cysteine, Zhuang et al. conjugated a charge tag (quaternary ammonium group) to cysteine through the click reaction, and then they used induced nESI MS to measure cysteine concentration in single living Hela and HepG2 cells.[73] Under the inspiration of bipolar spray process in induced nESI, Hu et al. developed a synchronized polarization induced electrospray ionization (SPI-ESI) method to acquire both positive-ion and negative-ion mass spectra in one measurement.[74]

Verbeck and coworkers developed an approach to coupling a nanomanipulator workstation with nano-ESI MS.[85] A solid quartz probe (tip size: ~8 μm) was used for cell puncture. Next, a Pd/Au-coated nanospray emitter (tip size: ~1 μm) was inserted into the cell through the puncture site to inject (using a precisely controlled four-channel pressure injector) a solvent mixture to extract metabolites and then draw the mixture for MS analysis. They observed a distinct difference of triacylglcerol abundance between the healthy and tumorous adipocytes, and then determined the heterogeneity of lipid droplets in living adipocyte cells.

A great effort has been put for quantitative single-cell MS analysis using techniques based on monochannel probes. Gholipour et al. used quartz capillary to fabricate the cell pressure probe, which was prefilled with oil mixture, to draw single-cell sap from tulip bulb (Fig. 4B).[71] Using a motorized micrometer, the sampling volume inside transparent probe can be precisely determined under a digital microscope. In addition, relative quantitation of metabolites in single cells can be measured by mixing internal reference solution (100 pl mannitol) with cell sap solution inside the capillary. In addition to syringe pump, electroosmotic extraction, which is induced by applying a voltage across the liquid/liquid interface, has been used to draw and inject small volumes (e.g., attoliter to picoliter) of solutions (Fig. 4C).[86] In the studies performed by Laskin and coworkers, one Pt electrode was inserted into a nanopipette (filled with a hydrophobic electrolyte solution), and the other was placed on the top of the cell.[87] A voltage was then applied on both electrodes to extract the cytoplasmic content into the nanopipette, which was later transferred to a mass spectrometer for analysis. Interestingly, the partition of metabolites between the aqueous and hydrophobic electrolyte solution can provide a separation of hydrophobic and hydrophilic analytes. In addition, glucose extracted from individual *A. cepa* epidermal cells was quantified by adding the internal standard compounds (glucose-d_2) into nanopipette using a second electroosmotic extraction.

To acquire richer structural information from metabolites in single cells, reactive SCMS studies have been carried out. The Yang group developed the micropipette needle, which was prepared by combing a pulled glass capillary needle with a fused silica capillary, to characterize carbon-carbon double (C=C) bonds in unsaturated lipids via Paternó-Büchi (PB) reactions (Fig. 4D).[75] Briefly, an organic solution (e.g., acetone or benzophenone dissolved in acetonitrile) was added into the micropipette needle and used as cell lysis solution and PB reagent. A whole suspension cell was drawn into the micropipette to produce lysate for MS analysis, and then PB reactions were initiated by UV irradiation for the same single-cell lysate. Finally, assisted by Python scripts, C=C bond locations were determined from mass spectra obtained before and after PB reactions.

Multichannel probes

Although a variety of different monochannel probes have been successfully used for SCMS metabolomics studies, experiments generally need to be conducted via two consecutive steps: single-cell sampling and MS analysis. To improve the experimental throughput, probes with multiple inner channels have been developed. The major techniques in this category include the nanospray DESI (nano-DESI),[88] single-probe,[89–91] and T-probe.[92, 93] The nano-DESI probe was assembled with a primary capillary and a secondary capillary, positioned at an optimum angle to each other. The analytes on the sample surface are desorbed into a flowing liquid bridge formed by the two capillaries. Although this method was originally designed for MS imaging studies,[94, 95] it has been used for SCMS experiments, in which the targets are individual human cheek cells placed on microscope glass.[88, 94, 95] In addition, by adding internal standard into the nano-DESI solvent, Bergman et al. used nano-DESI MS for online quantitative analysis of endogenous metabolites in single cell.[88]

The single-probe, developed by the Yang group, is a miniaturized, multifunctional sampling, and ionization device for real-time, in situ SCMS metabolomics studies (Fig. 5A).[89] A single-probe is produced by embedding a solvent-providing silica capillary and a nano-ESI emitter into a dual-bore quartz needle (tip size $<10\,\mu m$). The solvent-providing capillary, which is connected to a syringe pump, provides solvent to the dual-bore quartz needle tip, where solvent forms a liquid junction. By inserting the tip of dual-bore quartz needle into a target cell, metabolites can be extracted by solvent and then automatically delivered to the nano-ESI emitter for ionization. To maximize the amount of information in the positive ionization mode of single cells analysis, Pan et al. added dicationic ion-pairing reagents in the sampling solvent to convert the negatively charged metabolites into positively charged adducts.[96]

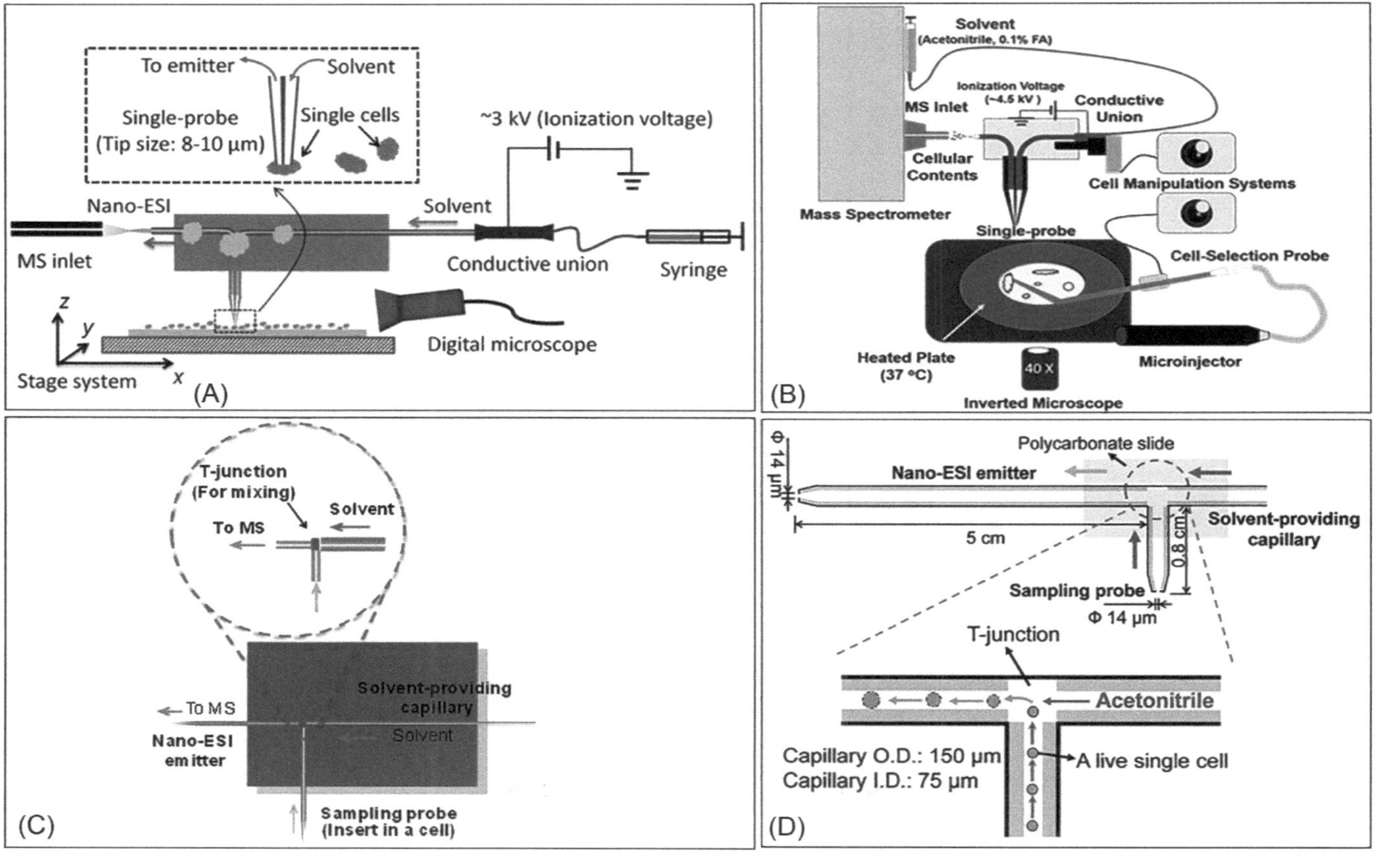

See the legend on next page.

Fig. 5 Examples of multichannel probes used in the probe-based SCMS metabolomics techniques. (A) Schematic of the Single-probe SCMS setup. A Single-probe was fabricated by embedding a solvent-providing capillary and a nano-ESI emitter in a laser-pulled dual-bore quartz needle. (B) Coupling the Single-probe MS to cell manipulation platform to quantify anticancer drug in single cancer cells isolated from patients. (C) Working mechanism of the T-probe. Three components (sampling probe, solvent-providing capillary, and nano-ESI emitter) were sandwiched by two polycarbonate slides. The suction force at the sampling probe tip was induced by nano-ESI and solvent flow. (D) The redesigned T-probe for the analysis of nonadherent single cells. A suspended single cell was captured by the sampling probe and drawn into the long nano-ESI emitter for rapid cell lysis and MS analysis. *((Panel A) Reproduced from reference Pan, N.; Rao, W.; Kothapalli, N. R.; Liu, R.; Burgett, A. W.; Yang, Z. The Single-Probe: a Miniaturized Multifunctional Device for Single Cell Mass Spectrometry Analysis. Anal. Chem.* ***2014****, 86 (19), 9376–9380, with permission from American Chemical Society. (Panel B) Reproduced from reference Bensen, R. C.; Standke, S. J.; Colby, D. H.; Kothapalli, N. R.; Le-McClain, A. T.; Patten, M. A.; Tripathi, A.; Heinlen, J. E.; Yang, Z.; Burgett, A. W. G. Single Cell Mass Spectrometry Quantification of Anticancer Drugs: Proof of Concept in Cancer Patients. ACS Pharmacol. Transl. Sci.* ***2021****, 4 (1), 96–100, with permission from American Chemical Society. (Panel C) Reproduced from reference Liu, R. M.; Pan, N.; Zhu, Y. L.; Yang, Z. B. T-Probe: An Integrated Microscale Device for Online in Situ Single Cell Analysis and Metabolic Profiling Using Mass Spectrometry. Anal. Chem.* ***2018****, 90 (18), 11078–11085, with permission from American Chemical Society. (Panel D) Reproduced from reference Zhu, Y. L.; Liu, R. M.; Yang, Z. B. Redesigning the T-Probe for Mass Spectrometry Analysis of Online Lysis of Non-adherent Single Cells. Anal. Chim. Acta* ***2019****, 1084, 53–59, with permission from Elsevier.)*

In addition, quantitative SCMS studies have been performed using this technique. To quantify drug uptake in single cells, Pan et al. added the internal standard into the sampling solvent, and used the Single-probe to analyze single cells cultured in microwells on glass chips.[90] To expand its capability of analyzing suspended cells, Standke et al. integrated the Single-probe with a cell manipulating platform.[91, 97] A single suspension cell can be captured a cell selection probe and transferred to the tip of the Single-probe for solvent induced lysis, followed by immediate MS analysis. By adding internal standard (^{15}N-gemcitabine) into the sampling solvent, this approach was used to measure intracellular drug (gemcitabine) concentrations in drug treated single cancer cells. Particularly, this technique allowed for quantifying amounts of gemcitabine in single cells isolated from patients undergoing gemcitabine treatment.[98] In addition to SCMS studies, the Single-probe device has been used to analyze extracellular metabolites inside live multicellular spheroids (Fig. 5B)[99] and high spatial resolution ambient MS imaging.[100–102]

The T-probe, another multichannel probe developed by the Yang group, has been used for adherence single-cell studies. Three components (i.e., the solvent-providing capillary, sampling probe, and nano-ESI emitter) produced from the fused silica capillary are connected at a T-junction and sandwiched by two polycarbonate substrates via thermo binding (Fig. 5C).[92] Once the sampling probe (tip size: ~5–8 μm) is inserted into a target single cell, the suction force, which is induced by nano-ESI and solvent flow, draws cellular contents toward the nano-ESI emitter for ionization. To analyze whole suspension cells, the redesigned T-probe has been developed. This new design has with two major changes: a sampling probe with larger orifice (~14 μm) and a longer nano-ESI emitter (5.5 cm). These two features allow for drawing a whole cell into the nano-ESI emitter for complete lysis prior to MS detection (Fig. 5D).[93]

4.2.2 Laser-based techniques

Laser-based techniques utilize laser for ablation sampling, whereas ionization of analytes occurs either simultaneously along with ablation [e.g., laser desorption/ionization droplet delivery (LDIDD)][103] or assisted by ESI [e.g., laser ablation electrospray ionization (LAESI)[104, 105]]. LDIDD uses pulsed UV laser beam to desorb and ionize analytes. Liquid droplets are sprayed onto the laser-irradiated region to minimize analyte loss and to assist the delivery of ions to a mass spectrometer inlet (Fig. 6A).[103] LDIDD MS achieved high sensitivity (e.g., the limit of detection of lysine is ~2 attomole) and spatial resolution (3 μm on mouse brain tissue). Using this method, different metabolomic profiles were observed between healthy and apoptotic HEK cells. LAESI, developed by Vertes and coworkers, utilizes a focused mid-infrared (IR) laser beam to ablate water-containing samples, and the ejected neutral species are then picked up and ionized by the charged droplets from ESI (Fig. 6B).[106] Its capability of single-cell studies was first demonstrated by analyzing metabolomic profiles of plant cells. To ablate single cells, the mid-IR laser pulses were delivered using an optical fiber with a sharp tip (radius of curvature ~15 μm).[104] More detailed studies were performed in later studies to improve the performance of this technique. For example, the effectiveness of IR laser delivery was evaluated using both the conventional and etched optical fibers.[105] Although the physical sizes of optical fiber tips are large compared with typical mammalian cells (e.g., diameter ~10 μm), this technique can be combined with microdissection (e.g., using a tungsten needle to expose the organelles and cytoplasm in plant cells) to study metabolites in subcellular compartment.[107] Because the microablation by focused laser beam has no significant influence on metabolites in neighboring cells, Shrestha et al. used this technique for MS imaging studies of cells in plant tissues.[108] The success of these studies lead to the development of automated

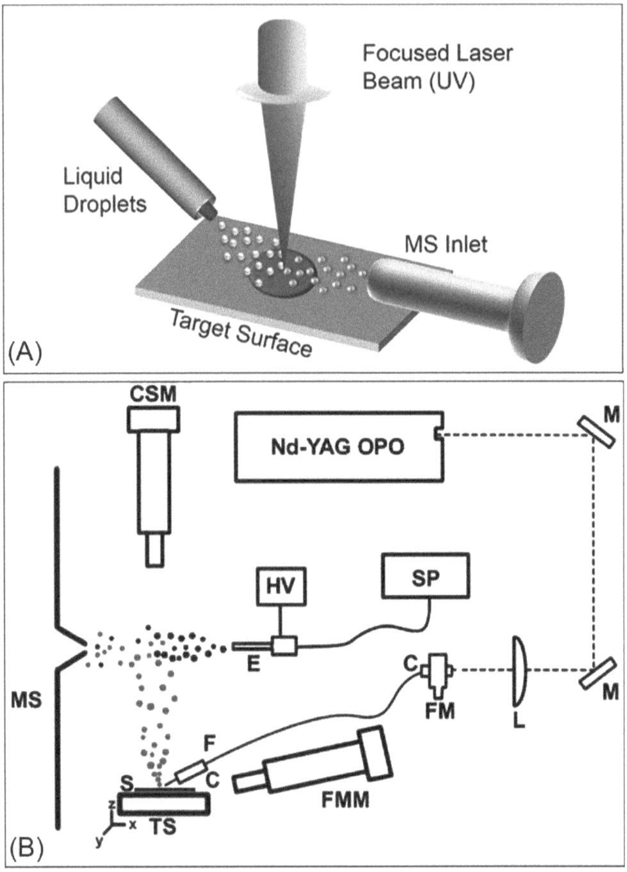

Fig. 6 Examples of laser-based SCMS metabolomics techniques. (A) Experimental setup of LDIDD-MS. Metabolite ions generated by UV laser were captured by liquid droplets and directed into MS. (B) Instrumental setup for LAESI-MS. Mid-IR laser was delivered to single cell through an etched optical fiber tip for ablating neutral species, which were picked and ionized by charged droplet from ESI. *((Panel A) Reproduced from reference Lee, J. K.; Jansson, E. T.; Nam, H. G.; Zare, R. N. High-Resolution Live-Cell Imaging and Analysis by Laser Desorption/Ionization Droplet Delivery Mass Spectrometry. Anal. Chem.* ***2016****, 88 (10), 5453–5461, with permission from American Chemical Society. (Panel B) Reproduced from reference Shrestha, B.; Vertes, A. In Situ Metabolic Profiling of Single Cells by Laser Ablation Electrospray Ionization Mass Spectrometry. Anal. Chem.* ***2009****, 81 (20), 8265–8271, with permission from American Chemical Society.)*

single-cell analysis for targeted morphologies.[109] Briefly, after the cell with particular morphology was recognized in the microscopy image, this cell was set as a target with centroid coordinates. A home-developed stage-control program was then utilized to control the translation stage, which moves the cell to the optical fiber for ablation and MS measurement.

4.3 Multimodal methods

As summarized in the above contexts, advancements in sampling and ionization methods play vital roles in promoting MS to be the primary tool for single-cell metabolomics studies. In fact, other analytical techniques, including separation, microfluidics, and spectroscopy, have been be integrated with MS for multidimensional analysis. These hybrid technologies provide additional structure information that can enhance molecular identification and promote experiment throughput.

4.3.1 Separation-assisted techniques

Most SCMS metabolomics experiments are carried out without separation, primarily due to two reasons: extremely limited amounts of analytes in individual cells and potential sample loss and dilution induced by separation. However, separation can greatly reduce the interference from matrix (i.e., increase detection sensitivity) and provide orthogonal structural information for molecular identification. According to the sequence to perform separation and ionization, there are major two types of separation methods, that is, preionization and postionization, used in ambient SCMS metabolomics studies.

As a typical preionization separation technique, capillary electrophoresis (CE), which utilizes electric field to separate analytes based on their charge to size ratio, can be used to provide a rapid separation of analytes with very small sample consumption. The combination of CE and MS, in which ESI is generally used as the ionization method, has been utilized in proteomics and metabolomics studies at the single-cell level. Sweedler and coworkers developed a CE-ESI setup, in which the ESI nebulizer was replaced by a coaxial sheath-flow interface, that can be coupled to MS (Fig. 7A).[110] During the experiment, a single cell (preserved in a home-built stainless steel nanovial) was injected from one end of the CE capillary. This capillary end was then placed into a stainless-steel containing background electrolyte, and CE separation was achieved by applying a voltage gradually increased from 0V to 20kV. This setup provided stable ion signal and high sensitivity (limit of detection was in the nanomolar range), allowing for the metabolic profiles to be obtained from single neuron cells and their subcellular regions. Nemes et al. improved the CE-ESI-MS design by using programed device to acquire a time-dependent output voltage.[111] They used this new method for qualitative and quantitative measurements of endogenous metabolites in neutrons.[112] They further modified the design to produce a CE-μESI

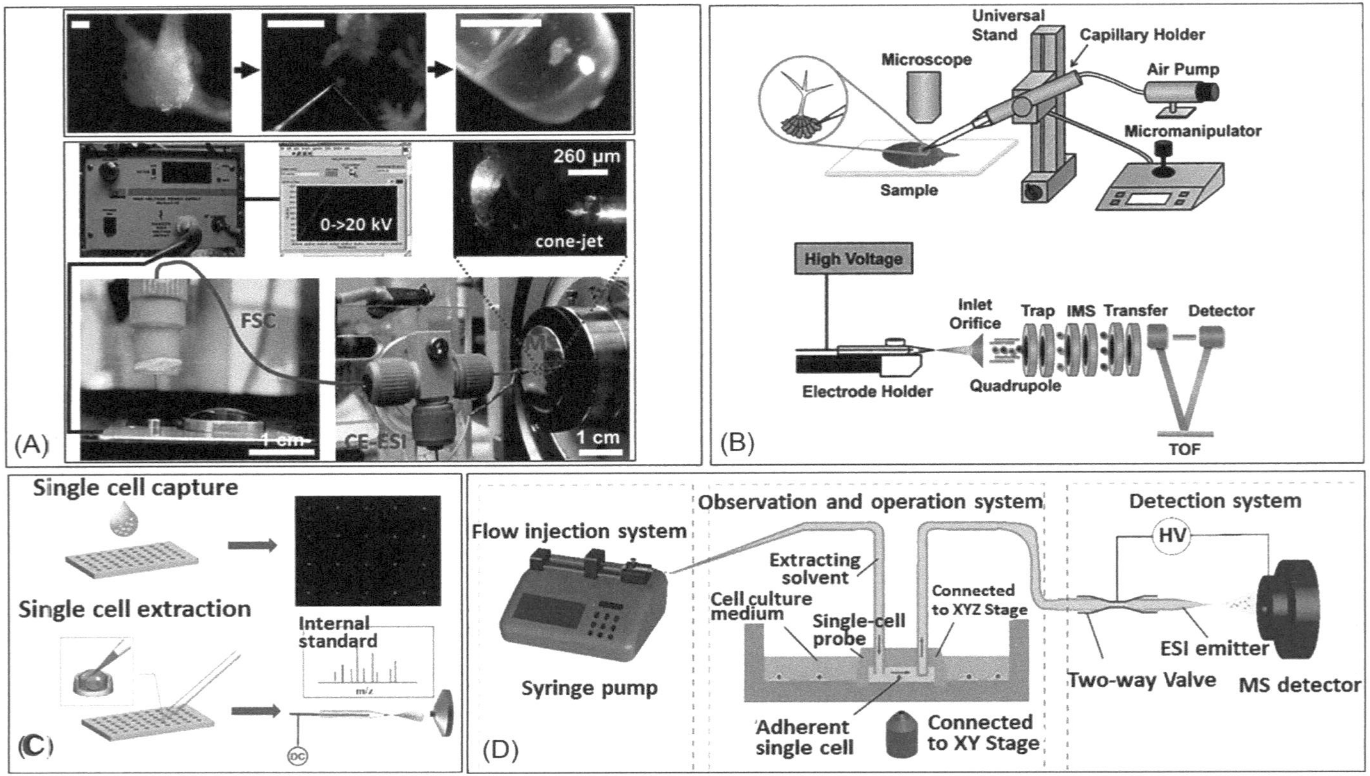

See the legend on opposite page.

Fig. 7 Examples of multimodal methods, including (A–B) separation-assisted and (C–D) microfluidic-based SCMS metabolomics techniques. (A) CE-ESI-MS platform for single-cell analysis. Single neutron cell lysate was introduced and separated by CE, and then cellular contents were sprayed into MS using a coaxial sheath-flow ESI source. (B) Experimental design of capillary microsampling ESI-IMS-MS. Separation of metabolite ions was achieved by ion mobility in studies of single plant and human cells. (C) Workflow of microwell-based nanoliter droplet microextraction and MS analysis of single cells. A microfluidic cell array was used to confine single cells, extraction solvent, and internal standard, allowing for quantitative measurement of glucose-phosphate concentration in single cells. (D) Working mechanism of microfluidic-based, in situ single-cell recognition system. Single-cell probe (fabricated from PDMS) was used for single-cell capture and cellular contents extraction from cells cultured in Petri dish. *((Panel A) Reproduced from reference Nemes, P.; Knolhoff, A. M.; Rubakhin, S. S.; Sweedler, J. V. Metabolic Differentiation of Neuronal Phenotypes by Single-Cell Capillary Electrophoresis-Electrospray Ionization-Mass Spectrometry. Anal. Chem. 2011, 83 (17), 6810–6817, with permission from American Chemical Society. (Panel B) Reproduced from reference Zhang, L. W.; Foreman, D. P.; Grant, P. A.; Shrestha, B.; Moody, S. A.; Villiers, F.; Kwake, J. M.; Vertes, A. In Situ Metabolic Analysis of Single Plant Cells by Capillary Microsampling and Electrospray Ionization Mass Spectrometry with Ion Mobility Separation. Analyst 2014, 139 (20), 5079–5085, with permission from Royal Society of Chemistry. (Panel C) Reproduced from reference Feng, J. X.; Zhang, X. C.; Huang, L.; Yao, H.; Yang, C. D.; Ma, X. X.; Zhang, S. C.; Zhang, X. R. Quantitation of Glucose-Phosphate in Single Cells by Microwell-Based Nanoliter Droplet Microextraction and Mass Spectrometry. Anal. Chem. 2019, 91 (9), 5613–5620, with permission from American Chemical Society. (Panel D) Reproduced from reference Huang, Q. S.; Mao, S. F.; Khan, M.; Li, W. W.; Zhang, Q.; Lin, J. M. Single-Cell Identification by Microfluidic-Based in Situ Extracting and Online Mass Spectrometric Analysis of Phospholipids Expression. Chem. Sci. 2020, 11 (1), 253–256, with permission from the author.*

interface, which was then used to compare and track metabolites in frog embryos during their development.[44, 45]

Separation can be also achieved in the gas phase after the ionization processes. Ion mobility spectrometry (IMS) separates ions based on their difference of mobilities, which are originated from differences in charge, mass, and cross section, in a buffer gas with the presence of an electric field. Vertes and coworker utilized the capillary microsampling for ESI-IMS-MS experiments (Fig. 7B).[113] The microcapillary, pulled from a glass capillary by a micropipette puller, was used to extract cellular contents via a motorized micromanipulator. After adding electrospray solution into the microcapillary, a platinum wire was inserted for applying voltage between the microcapillary and MS inlet. The generated ions were separated the IMS to eliminate the isobaric interferences and improve molecular coverage in MS analysis. This group further utilized this technique to study the influence of xenobiotic treatment on cellular metabolism (e.g., adenylate energy charge levels and lipid turnover rates) in single human hepatocellular carcinoma cells.[114]

4.3.2 Microfluidics-based techniques

The integration of SCMS with microfluidics added additional benefits for versatile studies. Particularly, microfluidics can be designed to achieve a variety of different functions, such as cell identification, sorting, and separation, in high-throughput analysis. So far, different types of microfluidic devices (e.g., microwells and droplet-based microfluidics) have been developed to isolate and capture single cells in suspension for analysis.[115] An integrated microfluidic device combining electrophoresis channel, in which rapid single-cell lysis and metabolites separation occur, and electrospray emitter has been developed, leading to a great throughput (12 cells per minute).[116] The combination of microfluidic chip and MALDI-MS achieved automatic, high-throughput analysis. In this design, a microwell-based microfluidic device was utilized to capture single cells for the formation of cell array, and MALDI-MS imaging was then used to analyze the metabolites in single cells automatically.[117] Jiaxin et al. further coupled the microwell-based microfluidic device with ESI-MS for metabolites quantitation analysis in single cells (Fig. 7C).[118] After single cells were captured in the microwells, a pulled capillary containing extraction solvent and internal standard was dip into microwells. The cellular contents along with the internal standard were then aspirated back to the capillary for MS detection. Using this method, author quantified concentrations of glucose phosphate in single K562 cells.

The integration of microfluidic device eliminates matrix effect when analyzing live cells present in culture medium or buffer that contains high concentrations of salts. Zhang et al. developed a three-phase droplet-based single-cell printing analysis system to separate and print single cells suspended in PBS (phosphate-buffered saline).[119] In this design, the organic extraction phase, octyl alcohol/acetonitrile, was used to extract cellular metabolites from the aqueous saline matrices. The separation of organic and aqueous phases occurred due to hydrophilic and hydrophobic interactions at the interface. The extract containing cellular metabolites was then printed to a microarray chip for MALDI-MS analysis. A high throughput (three to four single cells per second) has been achieved using this system. In another study, the Lin group integrated droplet-based inkjet printing, dielectrophoretic manipulation, and deemulsification interface with a microfluidic system to reduce the matrix interference for SCMS analysis.[120] Single cells were separated from the culture medium by splitting the droplet using a Y-shaped bifurcate structure in the dielectrophoretic channel. Using this system, they studied the heterogeneity of the healthy and cancer cells, and measured lipids' alterations due to drug treatment. In addition to suspension cells, microfluidic devices can be coupled with MS techniques for the analysis of adherent cells. The Lin group recently introduced a microfluidic-based in situ single-cell recognition system to achieve this purpose (Fig. 7D).[121] A single-cell probe fabricated from polydimethylsiloxane (PDMS) was used to capture and isolate single cells cultured in Petri dish with the assistance of a microscope. Methanol was then introduced to the compartment formed by the single-cell probe to extract lipids in a single cell, and the extraction solution was analyzed by MS. Authors conducted analyses of lipids (e.g., phosphatidylcholine) in single cells of different human cell lines.

4.3.3 Spectroscopy-assisted techniques

Spectroscopy techniques, including fluorescence and Raman, can be integrated for multidimensional analysis. In one of the early studies[122], fluorescence microscopy was used to locate fluorescently labeled cells in TOF-SIMS measurements of single-cell metabolomics analysis. To analyze changes in relative abundances of metabolites and lipids of cells in different mitotic stages (e.g., prometaphase, metaphase, anaphase, and cytokinesis), fluorescence microscopy (to determine mitotic stages) was combined with capillary microsampling ESI-IMS-MS to analyze single hepatocellular carcinoma cells.[123] Immunocytochemical assay is another analytical tool for classifying cell types, and this method has been combined with MALDI-MS to explore heterogeneous lipid compositions in single neurons and astrocytes.[124] Briefly, MALDI-MS was used to analyze rat brain cells present in mixed populations. After MS measurements, MALDI matrix was removed, cells were fixed, and immunocytochemical assay was subsequently performed to identify the neurons and astrocytes by labeling them with the corresponding antibodies.

Raman spectroscopy, as a nondestructive analysis tool, has been commonly used to acquire complementary chemical information (e.g., determination of functional groups in molecules). For single-cell dynamic studies, measurements using Raman spectroscopy and fluorescence imaging techniques were consecutively applied prior to MAIDI MS experiments.[125] In another study, these two spectroscopy techniques were integrated with matrix-free laser desorption/ionization MS for multidimensional chemical analysis of single algae cells,[14] which contain autofluorescent cellular contents.

5 Conclusion

A variety of analytical techniques, such as florescence, electrochemical analysis, MS, have demonstrated their versatile abilities for studies of metabolites in single cells, and remarkable advancements in this field have been achieved over the past decades. Benefitted by the development of instrumentation and innovative designs, highly sensitive, accurate, and efficient analyses have been accomplished, promoting our understanding of these biological systems with higher resolutions. With the impact of high-throughput analysis, the heterogeneity of genotype and phenotype can be rapidly revealed for large cell populations based on single-cell metabolomics. Although this short review only focuses on experimental technologies for metabolomics studies, it is expected that the integration of multiomics, such as transcriptomics and proteomics, and progress of new bioinformatics tools in the near future can greatly enhance single-cell analysis for fundamental research and biomedical applications.

Acknowledgment

This research project is partially supported by grants from National Institutes of Health (R01GM116116), National Science Foundation (OCE-1634630), and Research Council of the University of Oklahoma Norman Campus.

References

1. Strzelecka, P. M.; Ranzoni, A. M.; Cvejic, A. Dissecting Human Disease With Single-Cell Omics: Application in Model Systems and in the Clinic. *Dis. Model. Mech.* **2018**, *11* (11), 1–9.
2. Zenobi, R. Single-Cell Metabolomics: Analytical and Biological Perspectives. *Science* **2013**, *342* (6163), 1243259.
3. Zhang, L. W.; Vertes, A. Single-Cell Mass Spectrometry Approaches to Explore Cellular Heterogeneity. *Angew. Chem. Int. Ed.* **2018**, *57* (17), 4466–4477.
4. Kohnhorst, C. L.; Schmitt, D. L.; Sundaram, A.; An, S. O. Subcellular Functions of Proteins under Fluorescence Single-Cell Microscopy. *Biochim. Biophys. Acta Protein Proteomics* **2016**, *1864* (1), 77–84.
5. Liu, L. X.; Chen, D.Y.; Wang, J. B.; et al. Advances of Single-Cell Protein Analysis. *Cells* **2020**, *9* (5), 1–20.
6. Ohno, M.; Karagiannis, P.; Taniguchi, Y. Protein Expression Analyses at the Single Cell Level. *Molecules* **2014**, *19* (9), 13932–13947.
7. Cohen, D.; Dickerson, J. A.; Whitmore, C. D.; Turner, E. H.; Palcic, M. M.; Hindsgaul, O.; Dovichi, N. J. Chemical Cytometry: Fluorescence-Based Single-Cell Analysis. *Annu. Rev. Anal. Chem.* **2008**, *1*, 165–190.
8. Willingham, T. B.; Zhang, Y. F.; Andreoni, A.; Knutson, J. R.; Lee, D. Y.; Glancy, B. MitoRACE: Evaluating Mitochondrial Function In Vivo and in Single Cells with Subcellular Resolution Using Multiphoton NADH Autofluorescence. *J. Physiol. Lond.* **2019**, *597* (22), 5411–5428.
9. Yu, Q. R.; Heikal, A. A. Two-Photon Autofluorescence Dynamics Imaging Reveals Sensitivity of Intracellular NADH Concentration and Conformation to Cell Physiology at the Single-Cell Level. *J. Photochem. Photobiol. B* **2009**, *95* (1), 46–57.
10. Zhang, Z.; Milias-Argeitis, A.; Heinemann, M. Dynamic Single-Cell NAD(P)H Measurement Reveals Oscillatory Metabolism Throughout the *E. coli* Cell Division Cycle. *Sci. Rep.* **2018**, *8*.
11. Weber, P.; Wagner, M.; Kioschis, P.; et al. Tumor Cell Differentiation by Label-Free Fluorescence Microscopy. *J. Biomed. Opt.* **2012**, *17* (10), 1–5.
12. Heaster, T. M.; Walsh, A. J.; Zhao, Y.; et al. Autofluorescence Imaging Identifies Tumor Cell-Cycle Status on a Single-Cell Level. *J. Biophotonics* **2018**, *11* (1), 1–14.
13. An, G. H.; Suh, O. S.; Kwon, H. C.; Kim, K.; Johnson, E. A. Quantification of Carotenoids in Cells of Phaffia Rhodozyma by Autofluorescence. *Biotechnol. Lett.* **2000**, *22* (12), 1031–1034.
14. Urban, P. L.; Schmid, T.; Amantonico, A.; Zenobi, R. Multidimensional Analysis of Single Algal Cells by Integrating Microspectroscopy with Mass Spectrometry. *Anal. Chem.* **2011**, *83* (5), 1843–1849.
15. Kascakova, S.; Maigre, L.; Chevalier, J.; et al. Antibiotic Transport in Resistant Bacteria: Synchrotron UV Fluorescence Microscopy to Determine Antibiotic Accumulation with Single Cell Resolution. *PLoS ONE* **2012**, 7 (6), 1–9.
16. Anderson, A. B.; Gergen, J.; Arriaga, E. A. Detection of Doxorubicin and Metabolites in Cell Extracts and in Single Cells by Capillary Electrophoresis with Laser-Induced Fluorescence Detection. *J. Chromatogr. B* **2002**, *769* (1), 97–106.
17. Chacko, J. V.; Eliceiri, K. W. Autofluorescence Lifetime Imaging of Cellular Metabolism: Sensitivity Toward Cell Density, pH, Intracellular, and Intercellular Heterogeneity. *Cytometry A* **2019**, *95a* (1), 56–69.
18. Kasili, R. M.; Cullum, B. M.; Griffin, G. D.; Vo-Dinh, T. Nanosensor for In Vivo Measurement of the Carcinogen Benzo[a]Pyrene in a Single Cell. *J. Nanosci. Nanotechnol.* **2002**, *2* (6), 653–658.

19. Vo-Dinh, T.; Alarie, J. P.; Cullum, B. M.; Griffin, G. D. Antibody-Based Nanoprobe for Measurement of a Fluorescent Analyte in a Single Cell. *Nat. Biotechnol.* **2000**, *18* (7), 764–767.
20. Okumoto, S. Imaging Approach for Monitoring Cellular Metabolites and Ions Using Genetically Encoded Biosensors. *Curr. Opin. Biotechnol.* **2010**, *21* (1), 45–54.
21. Yaginuma, H.; Kawai, S.; Tabata, K. V.; Tomiyama, K.; Kakizuka, A.; Komatsuzaki, T.; Noji, H.; Imamura, H. Diversity in ATP Concentrations in a Single Bacterial Cell Population Revealed by Quantitative Single-Cell Imaging. *Sci. Rep.* **2014**, *4*, 6522.
22. Imamura, H.; Nhat, K. P. H.; Togawa, H.; Saito, K.; Iino, R.; Kato-Yamada, Y.; Nagai, T.; Noji, H. Visualization of ATP Levels inside Single Living Cells with Fluorescence Resonance Energy Transfer-Based Genetically Encoded Indicators. *Proc. Natl. Acad. Sci. U. S. A.* **2009**, *106* (37), 15651–15656.
23. Diaz-Garcia, C. M.; Lahmann, C.; Martinez-Francois, J. R.; Li, B. S.; Koveal, D.; Nathwani, N.; Rahman, M.; Keller, J. P.; Marvin, J. S.; Looger, L. L.; Yellen, G. Quantitative In Vivo Imaging of Neuronal Glucose Concentrations with a Genetically Encoded Fluorescence Lifetime Sensor. *J. Neurosci. Res.* **2019**, *97* (8), 946–960.
24. San Martin, A.; Ceballo, S.; Ruminot, I.; Lerchundi, R.; Frommer, W. B.; Barros, L. F. A Genetically Encoded FRET Lactate Sensor and its Use to Detect the Warburg Effect in Single Cancer Cells. *PLoS ONE* **2013**, *8* (2), e57712.
25. Bianchi-Smiraglia, A.; Rana, M. S.; Foley, C. E.; Paul, L. M.; Lipchick, B. C.; Moparthy, S.; Moparthy, K.; Fink, E. E.; Bagati, A.; Hurley, E.; Affronti, H. C.; Bakin, A. V.; Kandel, E. S.; Smiraglia, D. J.; Feltri, M. L.; Sousa, R.; Nikiforov, M. A. Internally Ratiometric Fluorescent Sensors for Evaluation of Intracellular GTP Levels and Distribution. *Nat. Methods* **2017**, *14* (10), 1003–1009.
26. Ueno, T.; Nagano, T. Fluorescent Probes for Sensing and Imaging. *Nat. Methods* **2011**, *8* (8), 642–645.
27. Jensen, E. C. Use of Fluorescent Probes: Their Effect on Cell Biology and Limitations. *Anat. Rec.* **2012**, *295* (12), 2031–2036.
28. Martinez-Zaguilan, R.; Tompkins, L.; Lynch, R. Simultaneous Analysis of Multiple Fluorescent Probes in Single Cells by Microspectroscopic Imaging. *SPIE* **1994**, *2137*, 17–28.
29. Joux, F.; Lebaron, P. Use of Fluorescent Probes to Assess Physiological Functions of Bacteria at Single-Cell Level. *Microbes Infect.* **2000**, *2* (12), 1523–1535.
30. Xue, M.; Wei, W.; Su, Y. P.; Kirn, J.; Shin, Y. S.; Mai, W. X.; Nathanson, D. A.; Heath, J. R. Chemical Methods for the Simultaneous Quantitation of Metabolites and Proteins from Single Cells. *J. Am. Chem. Soc.* **2015**, *137* (12), 4066–4069.
31. Li, Q. L.; Chen, P. L.; Fan, Y. Y.; Wang, X.; Xu, K. H.; Li, L.; Tang, B. Multicolor Fluorescence Detection-Based Microfluidic Device for Single-Cell Metabolomics: Simultaneous Quantitation of Multiple Small Molecules in Primary Liver Cells. *Anal. Chem.* **2016**, *88* (17), 8610–8616.
32. Bucher, E. S.; Wightman, R. M. Electrochemical Analysis of Neurotransmitters. *Annu. Rev. Anal. Chem.* **2015**, *8*, 239–261.
33. Schulte, A.; Schuhmann, W. Single-Cell Microelectrochemistry. *Angew. Chem. Int. Ed.* **2007**, *46* (46), 8760–8777.
34. Actis, P.; Mak, A. C.; Pourmand, N. Functionalized Nanopipettes: Toward Label-Free, Single Cell Biosensors. *Bioanal. Rev.* **2010**, *1* (2–4), 177–185.
35. Li, X. C.; Majdi, S.; Dunevall, J.; Fathali, H.; Ewing, A. G. Quantitative Measurement of Transmitters in Individual Vesicles in the Cytoplasm of Single Cells with Nanotip Electrodes. *Angew. Chem. Int. Ed.* **2015**, *54* (41), 11978–11982.
36. Liu, Y. Z.; Li, M. N.; Zhang, F.; Zhu, A. W.; Shi, G. Y. Development of Au Disk Nanoelectrode Down to 3 nm in Radius for Detection of Dopamine Release from a Single Cell. *Anal. Chem.* **2015**, *87* (11), 5531–5538.

37. Abe, T.; Lau, Y.Y.; Ewing, A. G. Characterization of Glucose Microsensors for Intracellular Measurements. *Anal. Chem.* **1992**, *64* (18), 2160–2163.
38. Wang, Y.; Fathali, H.; Mishra, D.; Olsson, T.; Keighron, J. D.; Skibicka, K. P.; Cans, A. S. Counting the Number of Glutamate Molecules in Single Synaptic Vesicles. *J. Am. Chem. Soc.* **2019**, *141* (44), 17507–17511.
39. Malinski, T.; Taha, Z. Nitric Oxide Release from a Single Cell Measured In Situ by a Porphyrinic-Based Microsensor. *Nature* **1992**, *358* (6388), 676–678.
40. Wang, X.; Wang, J.; Fu, H.; Liu, D.; Chen, Z. Determination of Glutathione in Single HepG2 Cells by Capillary Electrophoresis with Reduced Graphene Oxide Modified Microelectrode. *Electrophoresis* **2014**, *35* (23), 3371–3378.
41. Xia, F. Q.; Jin, W. R.; Yin, X. F.; Fang, Z. L. Single-Cell Analysis by Electrochemical Detection with a Microfluidic Device. *J. Chromatogr. A* **2005**, *1063* (1–2), 227–233.
42. Pan, R.; Xu, M.; Jiang, D.; Burgess, J. D.; Chen, H.Y. Nanokit for Single-Cell Electrochemical Analyses. *Proc. Natl. Acad. Sci. U. S. A.* **2016**, *113* (41), 11436–11440.
43. Arnold, R. J.; Reilly, J. P. Fingerprint Matching of *E-coli* Strains with Matrix-Assisted Laser Desorption Ionization Time-of-Flight Mass Spectrometry of Whole Cells Using a Modified Correlation Approach. *Rapid Commun. Mass Spectrom.* **1998**, *12* (10), 630–636.
44. Onjiko, R. M.; Morris, S. E.; Moody, S. A.; Nemes, P. Single-Cell Mass Spectrometry with Multi-Solvent Extraction Identifies Metabolic Differences between Left and Right Blastomeres in the 8-Cell Frog (Xenopus) Embryo. *Analyst* **2016**, *141* (12), 3648–3656.
45. Onjiko, R. M.; Moody, S. A.; Nemes, P. Single-Cell Mass Spectrometry Reveals Small Molecules that Affect Cell Fates in the 16-Cell Embryo. *Proc. Natl. Acad. Sci. U. S. A.* **2015**, *112* (21), 6545–6550.
46. Colliver, T. L.; Brummel, C. L.; Pacholski, M. L.; Swanek, F. D.; Ewing, A. G.; Winograd, N. Atomic and Molecular Imaging at the Single-Cell Level with TOF-SIMS. *Anal. Chem.* **1997**, *69* (13), 2225–2231.
47. Mai, F. D.; Chen, B. J.; Wu, L. C.; Li, F.Y.; Chen, W. K. Imaging of Single Liver Tumor Cells Intoxicated by Heavy Metals Using ToF-SIMS. *Appl. Surf. Sci.* **2006**, *252* (19), 6809–6812.
48. Li, H. W.; Hua, X.; Long, Y. T. Graphene Quantum Dots Enhanced ToF-SIMS for Single-Cell Imaging. *Anal. Bioanal. Chem.* **2019**, *411* (18), 4025–4030.
49. Do, T. D.; Comi, T. J.; Dunham, S. J. B.; Rubakhin, S. S.; Sweedler, J.V. Single Cell Profiling Using Ionic Liquid Matrix-Enhanced Secondary Ion Mass Spectrometry for Neuronal Cell Type Differentiation. *Anal. Chem.* **2017**, *89* (5), 3078–3086.
50. Altelaar, A. F. M.; Klinkert, I.; Jalink, K.; de Lange, R. P. J.; Adan, R. A. H.; Heeren, R. M. A.; Piersma, S. R. Gold-Enhanced Biomolecular Surface Imaging of Cells and Tissue by SIMS and MALDI Mass Spectrometry. *Anal. Chem.* **2006**, *78* (3), 734–742.
51. Wehrli, P. M.; Lindberg, E.; Angerer, T. B.; Wold, A. E.; Gottfries, J.; Fletcher, J. S. Maximising the Potential for Bacterial Phenotyping Using Time-of-Flight Secondary Ion Mass Spectrometry with Multivariate Analysis and Tandem Mass Spectrometry. *Surf. Interface Anal.* **2014**, *46*, 173–176.
52. Robinson, M. A.; Graham, D. J.; Morrish, F.; Hockenbery, D.; Gamble, L. J. Lipid Analysis of Eight Human Breast Cancer Cell Lines with ToF-SIMS. *Biointerphases* **2016**, *11* (2), 02A303.
53. Ostrowski, S. G.; Kurczy, M. E.; Roddy, T. P.; Winograd, N.; Ewing, A. G. Secondary Ion MS Imaging to Relatively Quantify Cholesterol in the Membranes of Individual Cells from Differentially Treated Populations. *Anal. Chem.* **2007**, *79* (10), 3554–3560.
54. Rabbani, S.; Fletcher, J. S.; Lockyer, N. P.; Vickerman, J. C. Exploring Subcellular Imaging on the Buncher-ToF J105 3D Chemical Imager. *Surf. Interface Anal.* **2011**, *43* (1–2), 380–384.

55. Jiang, H.; Favaro, E.; Goulbourne, C. N.; Rakowska, P. D.; Hughes, G. M.; Ryadnov, M. G.; Fong, L. G.; Young, S. G.; Ferguson, D. J.; Harris, A. L.; Grovenor, C. R. Stable Isotope Imaging of Biological Samples with High Resolution Secondary Ion Mass Spectrometry and Complementary Techniques. *Methods* **2014**, *68* (2), 317–324.
56. Ong, T. H.; Kissick, D. J.; Jansson, E. T.; Comi, T. J.; Romanova, E. V.; Rubakhin, S. S.; Sweedler, J. V. Classification of Large Cellular Populations and Discovery of Rare Cells Using Single Cell Matrix-Assisted Laser Desorption/Ionization Time-of-Flight Mass Spectrometry. *Anal. Chem.* **2015**, *87* (14), 7036–7042.
57. Ibanez, A. J.; Fagerer, S. R.; Schmidt, A. M.; Urban, P. L.; Jefimovs, K.; Geiger, P.; Dechant, R.; Heinemann, M.; Zenobi, R. Mass Spectrometry-Based Metabolomics of Single Yeast Cells. *Proc. Natl. Acad. Sci. U. S. A.* **2013**, *110* (22), 8790–8794.
58. Urban, P. L.; Jefimovs, K.; Amantonico, A.; Fagerer, S. R.; Schmid, T.; Madler, S.; Puigmarti-Luis, J.; Goedecke, N.; Zenobi, R. High-Density Micro-Arrays for Mass Spectrometry. *Lab Chip* **2010**, *10* (23), 3206–3209.
59. Kruse, R. A.; Rubakhin, S. S.; Romanova, E. V.; Bohn, P. W.; Sweedler, J. V. Direct Assay of Aplysia Tissues and Cells with Laser Desorption/Ionization Mass Spectrometry on Porous Silicon. *J. Mass Spectrom.* **2001**, *36* (12), 1317–1322.
60. Greving, M. P.; Patti, G. J.; Siuzdak, G. Nanostructure-Initiator Mass Spectrometry Metabolite Analysis and Imaging. *Anal. Chem.* **2011**, *83* (1), 2–7.
61. O'Brien, P. J.; Lee, M.; Spilker, M. E.; Zhang, C. C.; Yan, Z.; Nichols, T. C.; Li, W.; Johnson, C. H.; Patti, G. J.; Siuzdak, G. Monitoring Metabolic Responses to Chemotherapy in Single Cells and Tumors Using Nanostructure-Initiator Mass Spectrometry (NIMS) Imaging. *Cancer Metabol.* **2013**, *1* (1), 4.
62. Walker, B. N.; Stolee, J. A.; Vertes, A. Nanophotonic Ionization for Ultratrace and Single-Cell Analysis by Mass Spectrometry. *Anal. Chem.* **2012**, *84* (18), 7756–7762.
63. Chen, Y. T.; Li, G. Y.; Yuan, S. M.; Pan, Y.; Liu, Y. Z.; Huang, G. M. Ultrafast Microelectrophoresis: Behind Direct Mass Spectrometry Measurements of Proteins and Metabolites in Living Cell/Cells. *Anal. Chem.* **2019**, *91* (16), 10441–10447.
64. Cooks, R. G.; Ouyang, Z.; Takats, Z.; Wiseman, J. M. Ambient Mass Spectrometry. *Science* **2006**, *311* (5767), 1566–1570.
65. Hiraoka, K.; Nishidate, K.; Mori, K.; Asakawa, D.; Suzuki, S. Development of Probe Electrospray Using a Solid Needle. *Rapid Commun. Mass Spectrom.* **2007**, *21* (18), 3139–3144.
66. Arthur, C. L.; Pawliszyn, J. Solid-Phase Microextraction with Thermal-Desorption Using Fused-Silica Optical Fibers. *Anal. Chem.* **1990**, *62* (19), 2145–2148.
67. Gong, X. Y.; Zhao, Y. Y.; Cai, S. Q.; Fu, S. J.; Yang, C. D.; Zhang, S. C.; Zhang, X. R. Single Cell Analysis with Probe ESI-Mass Spectrometry: Detection of Metabolites at Cellular and Subcellular Levels. *Anal. Chem.* **2014**, *86* (8), 3809–3816.
68. Yu, Z.; Chen, L. C.; Ninomiya, S.; Mandal, M. K.; Hiraoka, K.; Nonami, H. Piezoelectric Inkjet Assisted Rapid Electrospray Ionization Mass Spectrometric Analysis of Metabolites in Plant Single Cells Via a Direct Sampling Probe. *Analyst* **2014**, *139* (22), 5734–5739.
69. Chen, F. M.; Lin, L. Y.; Zhang, J.; He, Z. Y.; Uchiyama, K.; Lin, J. M. Single-Cell Analysis Using Drop-on-Demand Inkjet Printing and Probe Electrospray Ionization Mass Spectrometry. *Anal. Chem.* **2016**, *88* (8), 4354–4360.
70. Zheng, Y.; Liu, Z. Q.; Xing, J. P.; Zheng, Z.; Pi, Z. F.; Song, F. R.; Liu, S. In Situ Analysis of Single Cell and Biological Samples with rGO-Cu Functional Probe ESI-MS Spectrometry. *Talanta* **2020**, *211*, 120751.
71. Gholipour, Y.; Erra-Balsells, R.; Hiraoka, K.; Nonami, H. Living Cell Manipulation, Manageable Sampling, and Shotgun Picoliter Electrospray Mass Spectrometry for Profiling Metabolites. *Anal. Biochem.* **2013**, *433* (1), 70–78.
72. Tsuyama, N.; Mizuno, H.; Tokunaga, E.; Masujima, T. Live Single-Cell Molecular Analysis by Video-Mass Spectrometry. *Anal. Sci.* **2008**, *24* (5), 559–561.

73. Zhuang, M. H.; Hou, Z. H.; Chen, P. Y.; Liang, G. L.; Huang, G. M. Introducing Charge Tag Via Click Reaction in Living Cells for Single Cell Mass Spectrometry. *Chem. Sci.* **2020**, *11* (28), 7308–7312.
74. Hu, J.; Jiang, X.-X.; Wang, J.; Guan, Q.-Y.; Zhang, P.-K.; Xu, J.-J.; Chen, H.-Y. Synchronized Polarization Induced Electrospray: Comprehensively Profiling Biomolecules in Single Cells by Combining Both Positive-Ion and Negative-Ion Mass Spectra. *Anal. Chem.* **2016**, *88* (14), 7245–7251.
75. Zhu, Y. L.; Wang, W. H.; Yang, Z. B. Combining Mass Spectrometry with Paterno-Buchi Reaction to Determine Double-Bond Positions in Lipids at the Single-Cell Level. *Anal. Chem.* **2020**, *92* (16), 11380–11387.
76. Nakashima, T.; Wada, H.; Morita, S.; Erra-Balsells, R.; Hiraoka, K.; Nonami, H. Single-Cell Metabolite Profiling of Stalk and Glandular Cells of Intact Trichomes with Internal Electrode Capillary Pressure Probe Electrospray Ionization Mass Spectrometry. *Anal. Chem.* **2016**, *88* (6), 3049–3057.
77. Abouleila, Y.; Onidani, K.; Ali, A.; Shoji, H.; Kawai, T.; Lim, C. T.; Kumar, V.; Okaya, S.; Kato, K.; Hiyama, E.; Yanagida, T.; Masujima, T.; Shimizu, Y.; Honda, K. Live Single Cell Mass Spectrometry Reveals Cancer-Specific Metabolic Profiles of Circulating Tumor Cells. *Cancer Sci.* **2019**, *110* (2), 697–706.
78. Fujii, T.; Matsuda, S.; Tejedor, M. L.; Esaki, T.; Sakane, I.; Mizuno, H.; Tsuyama, N.; Masujima, T. Direct Metabolomics for Plant Cells by Live Single-Cell Mass Spectrometry. *Nat. Protoc.* **2015**, *10* (9), 1445–1456.
79. Masujima, T. Live Single-Cell Mass Spectrometry. *Anal. Sci.* **2009**, *25* (8), 953–960.
80. Hiyama, E.; Ali, A.; Amer, S.; Harada, T.; Shimamoto, K.; Furushima, R.; Abouleila, Y.; Emara, S.; Masujima, T. Direct Lipido-Metabolomics of Single Floating Cells for Analysis of Circulating Tumor Cells by Live Single-Cell Mass Spectrometry. *Anal. Sci.* **2015**, *31* (12), 1215–1217.
81. Esaki, T.; Masujima, T. Fluorescence Probing Live Single-Cell Mass Spectrometry for Direct Analysis of Organelle Metabolism. *Anal. Sci.* **2015**, *31* (12), 1211–1213.
82. Ali, A.; Abouleila, Y.; Amer, S.; Furushima, R.; Emara, S.; Equis, S.; Cotte, Y.; Masujima, T. Quantitative Live Single-Cell Mass Spectrometry with Spatial Evaluation by Three-Dimensional Holographic and Tomographic Laser Microscopy. *Anal. Sci.* **2016**, *32* (2), 125–127.
83. Zhao, J. B.; Zhang, F.; Guo, Y. L. Quantitative Analysis of Metabolites at the Single-Cell Level by Hydrogen Flame Desorption Ionization Mass Spectrometry. *Anal. Chem.* **2019**, *91* (4), 2752–2758.
84. Huang, G. M.; Li, G. T.; Cooks, R. G. Induced Nanoelectrospray Ionization for Matrix-Tolerant and High-Throughput Mass Spectrometry. *Angew. Chem. Int. Ed.* **2011**, *50* (42), 9907–9910.
85. Phelps, M.; Hamilton, J.; Verbeck, G. F. Nanomanipulation-Coupled Nanospray Mass Spectrometry as an Approach for Single Cell Analysis. *Rev. Sci. Instrum.* **2014**, *85* (12), 124101.
86. Laforge, F. O.; Carpino, J.; Rotenberg, S. A.; Mirkin, M. V. Electrochemical Attosyringe. *Proc. Natl. Acad. Sci. U. S. A.* **2007**, *104* (29), 11895–11900.
87. Yin, R. C.; Prabhakaran, V.; Laskin, J. Quantitative Extraction and Mass Spectrometry Analysis at a Single-Cell Level. *Anal. Chem.* **2018**, *90* (13), 7937–7945.
88. Bergman, H. M.; Lanekoff, I. Profiling and Quantifying Endogenous Molecules in Single Cells Using Nano-DESI MS. *Analyst* **2017**, *142* (19), 3639–3647.
89. Pan, N.; Rao, W.; Kothapalli, N. R.; Liu, R.; Burgett, A. W.; Yang, Z. The Single-Probe: a Miniaturized Multifunctional Device for Single Cell Mass Spectrometry Analysis. *Anal. Chem.* **2014**, *86* (19), 9376–9380.

90. Pan, N.; Standke, S. J.; Kothapalli, N. R.; Sun, M.; Bensen, R. C.; Burgett, A. W. G.; Yang, Z. B. Quantification of Drug Molecules in Live Single Cells Using the Single-Probe Mass Spectrometry Technique. *Anal. Chem.* **2019**, *91* (14), 9018–9024.
91. Standke, S. J.; Colby, D. H.; Bensen, R. C.; Burgett, A. W. G.; Yang, Z. Mass Spectrometry Measurement of Single Suspended Cells Using a Combined Cell Manipulation System and a Single-Probe Device. *Anal. Chem.* **2019**, *91* (3), 1738–1742.
92. Liu, R. M.; Pan, N.; Zhu, Y. L.; Yang, Z. B. T-Probe: An Integrated Microscale Device for Online in Situ Single Cell Analysis and Metabolic Profiling Using Mass Spectrometry. *Anal. Chem.* **2018**, *90* (18), 11078–11085.
93. Zhu, Y. L.; Liu, R. M.; Yang, Z. B. Redesigning the T-Probe for Mass Spectrometry Analysis of Online Lysis of Non-adherent Single Cells. *Anal. Chim. Acta* **2019**, *1084*, 53–59.
94. Yin, R. C.; Burnum-Johnson, K. E.; Sun, X. F.; Dey, S. K.; Laskin, J. High Spatial Resolution Imaging of Biological Tissues Using Nanospray Desorption Electrospray Ionization Mass Spectrometry. *Nat. Protoc.* **2019**, *14* (12), 3445–3470.
95. Laskin, J.; Heath, B. S.; Roach, P. J.; Cazares, L.; Semmes, O. J. Tissue Imaging Using Nanospray Desorption Electrospray Ionization Mass Spectrometry. *Anal. Chem.* **2012**, *84* (1), 141–148.
96. Pan, N.; Rao, W.; Standke, S. J.; Yang, Z. B. Using Dicationic Ion-Pairing Compounds to Enhance the Single Cell Mass Spectrometry Analysis Using the Single-Probe: A Microscale Sampling and Ionization Device. *Anal. Chem.* **2016**, *88* (13), 6812–6819.
97. Standke, S. J.; Colby, D. H.; Bensen, R. C.; Burgett, A. W. G.; Yang, Z. B. Integrated Cell Manipulation Platform Coupled with the Single-Probe for Mass Spectrometry Analysis of Drugs and Metabolites in Single Suspension Cells. *J. Vis. Exp.* **2019**, *148*, e59875.
98. Bensen, R. C.; Standke, S. J.; Colby, D. H.; Kothapalli, N. R.; Le-McClain, A. T.; Patten, M. A.; Tripathi, A.; Heinlen, J. E.; Yang, Z.; Burgett, A. W. G. Single Cell Mass Spectrometry Quantification of Anticancer Drugs: Proof of Concept in Cancer Patients. *ACS Pharmacol. Transl. Sci.* **2021**, *4* (1), 96–100.
99. Sun, M.; Tian, X.; Yang, Z. B. Microscale Mass Spectrometry Analysis of Extracellular Metabolites in Live Multicellular Tumor Spheroids. *Anal. Chem.* **2017**, *89* (17), 9069–9076.
100. Tian, X.; Zhang, G. W.; Zou, Z.; Yang, Z. B. Anticancer Drug Affects Metabolomic Profiles in Multicellular Spheroids: Studies Using Mass Spectrometry Imaging Combined with Machine Learning. *Anal. Chem.* **2019**, *91* (9), 5802–5809.
101. Tian, X.; Zhang, G. W.; Shao, Y. H.; Yang, Z. B. Towards Enhanced Metabolomic Data Analysis of Mass Spectrometry Image: Multivariate Curve Resolution and Machine Learning. *Anal. Chim. Acta* **2018**, *1037*, 211–219.
102. Rao, W.; Pan, N.; Yang, Z. B. High Resolution Tissue Imaging Using the Single-Probe Mass Spectrometry under Ambient Conditions. *J. Am. Soc. Mass Spectrom.* **2015**, *26* (6), 986–993.
103. Lee, J. K.; Jansson, E. T.; Nam, H. G.; Zare, R. N. High-Resolution Live-Cell Imaging and Analysis by Laser Desorption/Ionization Droplet Delivery Mass Spectrometry. *Anal. Chem.* **2016**, *88* (10), 5453–5461.
104. Shrestha, B.; Vertes, A. In Situ Metabolic Profiling of Single Cells by Laser Ablation Electrospray Ionization Mass Spectrometry. *Anal. Chem.* **2009**, *81* (20), 8265–8271.
105. Shrestha, B.; Nemes, P.; Vertes, A. Ablation and Analysis of Small Cell Populations and Single Cells by Consecutive Laser Pulses. *Appl. Phys. A Mater. Sci. Process.* **2010**, *101* (1), 121–126.
106. Nemes, P.; Vertes, A. Laser Ablation Electrospray Ionization for Atmospheric Pressure, in Vivo, and Imaging Mass Spectrometry. *Anal. Chem.* **2007**, *79* (21), 8098–8106.

107. Stolee, J. A.; Shrestha, B.; Mengistu, G.; Vertes, A. Observation of Subcellular Metabolite Gradients in Single Cells by Laser Ablation Electrospray Ionization Mass Spectrometry. *Angew. Chem. Int. Ed.* **2012**, *51* (41), 10386–10389.
108. Shrestha, B.; Patt, J. M.; Vertes, A. In Situ Cell-by-Cell Imaging and Analysis of Small Cell Populations by Mass Spectrometry. *Anal. Chem.* **2011**, *83* (8), 2947–2955.
109. Li, H.; Smith, B. K.; Shrestha, B.; Márk, L.; Vertes, A. Automated Cell-by-Cell Tissue Imaging and Single-Cell Analysis for Targeted Morphologies by Laser Ablation Electrospray Ionization Mass Spectrometry. In *Mass Spectrometry Imaging of Small Molecules*; Lin, H., Ed.; Springer New York: New York, NY, 2015; pp. 117–127.
110. Lapainis, T.; Rubakhin, S. S.; Sweedler, J.V. Capillary Electrophoresis with Electrospray Ionization Mass Spectrometric Detection for Single-Cell Metabolomics. *Anal. Chem.* **2009**, *81* (14), 5858–5864.
111. Nemes, P.; Knolhoff, A. M.; Rubakhin, S. S.; Sweedler, J. V. Metabolic Differentiation of Neuronal Phenotypes by Single-Cell Capillary Electrophoresis-Electrospray Ionization-Mass Spectrometry. *Anal. Chem.* **2011**, *83* (17), 6810–6817.
112. Nemes, P.; Rubakhin, S. S.; Aerts, J. T.; Sweedler, J.V. Qualitative and Quantitative Metabolomic Investigation of Single Neurons by Capillary Electrophoresis Electrospray Ionization Mass Spectrometry. *Nat. Protoc.* **2013**, *8* (4), 783–799.
113. Zhang, L. W.; Foreman, D. P.; Grant, P. A.; Shrestha, B.; Moody, S. A.; Villiers, F.; Kwake, J. M.; Vertes, A. In Situ Metabolic Analysis of Single Plant Cells by Capillary Microsampling and Electrospray Ionization Mass Spectrometry with Ion Mobility Separation. *Analyst* **2014**, *139* (20), 5079–5085.
114. Zhang, L. W.; Vertes, A. Energy Charge, Redox State, and Metabolite Turnover in Single Human Hepatocytes Revealed by Capillary Microsampling Mass Spectrometry. *Anal. Chem.* **2015**, *87* (20), 10397–10405.
115. Feng, D. S.; Xu, T. R.; Li, H.; Shi, X. Z.; Xu, G. W. Single-Cell Metabolomics Analysis by Microfluidics and Mass Spectrometry: Recent New Advances. *J. Anal. Test* **2020**, *4* (3), 198–209.
116. Mellors, J. S.; Jorabchi, K.; Smith, L. M.; Ramsey, J. M. Integrated Microfluidic Device for Automated Single Cell Analysis Using Electrophoretic Separation and Electrospray Ionization Mass Spectrometry. *Anal. Chem.* **2010**, *82* (3), 967–973.
117. Xie, W. Y.; Gao, D.; Jin, F.; Jiang, Y. Y.; Liu, H. X. Study of Phospholipids in Single Cells Using an Integrated Microfluidic Device Combined with Matrix-Assisted Laser Desorption/Ionization Mass Spectrometry. *Anal. Chem.* **2015**, *87* (14), 7052–7059.
118. Feng, J. X.; Zhang, X. C.; Huang, L.; Yao, H.; Yang, C. D.; Ma, X. X.; Zhang, S. C.; Zhang, X. R. Quantitation of Glucose-Phosphate in Single Cells by Microwell-Based Nanoliter Droplet Microextraction and Mass Spectrometry. *Anal. Chem.* **2019**, *91* (9), 5613–5620.
119. Li, Q.; Tang, F.; Huo, X. M.; Huang, X.; Zhang, Y.; Wang, X. H.; Zhang, X. R. Native State Single-Cell Printing System and Analysis for Matrix Effects. *Anal. Chem.* **2019**, *91* (13), 8115–8122.
120. Zhang, W. F.; Li, N.; Lin, L.; Huang, Q. S.; Uchiyama, K.; Lin, J. M. Concentrating Single Cells in Picoliter Droplets for Phospholipid Profiling on a Microfluidic System. *Small* **2020**, *16* (9), 1903402.
121. Huang, Q. S.; Mao, S. F.; Khan, M.; Li, W. W.; Zhang, Q.; Lin, J. M. Single-Cell Identification by Microfluidic-Based in Situ Extracting and Online Mass Spectrometric Analysis of Phospholipids Expression. *Chem. Sci.* **2020**, *11* (1), 253–256.
122. Roddy, T. P.; Cannon, D. M.; Meserole, C. A.; Winograd, N.; Ewing, A. G. Imaging of Freeze-Fractured Cells with in Situ Fluorescence and Time-of-Flight Secondary Ion Mass Spectrometry. *Anal. Chem.* **2002**, *74* (16), 4011–4019.

123. Zhang, L. W.; Sevinsky, C. J.; Davis, B. M.; Vertes, A. Single-Cell Mass Spectrometry of Subpopulations Selected by Fluorescence Microscopy. *Anal. Chem.* **2018**, *90* (7), 4626–4634.
124. Neumann, E. K.; Comi, T. J.; Rubakhin, S. S.; Sweedler, J. V. Lipid Heterogeneity between Astrocytes and Neurons Revealed by Single-Cell MALDI-MS Combined with Immunocytochemical Classification. *Angew. Chem. Int. Ed.* **2019**, *58* (18), 5910–5914.
125. Fagerer, S. R.; Schmid, T.; Ibanez, A. J.; Pabst, M.; Steinhoff, R.; Jefimovs, K.; Urban, P. L.; Zenobi, R. Analysis of Single Algal Cells by Combining Mass Spectrometry with Raman and Fluorescence Mapping. *Analyst* **2013**, *138* (22), 6732–6736.

CHAPTER 4

Single-cell electrical characterization

Jonathan Cottet[a,*] and Federica Caselli[b,*]
[a]Department of Mechanical Engineering, Massachusetts Institute of Technology, Cambridge, MA, United States
[b]Department of Civil Engineering and Computer Science, University of Rome Tor Vergata, Rome, Italy

1 Introduction

Interactions between electric fields and biological cells have been of interest for well over a century.[1–4] Electrical properties of living cells play significant roles in understanding of various biological activities including disease progression at the cellular and molecular levels.[5] Moreover, electric fields offer tremendous opportunities for analysis and manipulation of biological cells, by means of versatile, effective, and yet simple sensing and actuation features. Since 1990, the introduction of microfluidics and the development of key enabling technologies (microelectronics, microfabrication, and nanofabrication) boosted the progress in this field of study, and the reasons are many. Cell size is in the order of microns and therefore microfluidic devices represent the perfect tools for cell manipulation and analysis, especially if single-cell level is pursued. Electrical sensing has the key advantage of being label-free, electrical forces scale favorably with miniaturization, and electrical systems are at the forefront of integration, adaptability, and portability for diverse applications. This combination of electric fields and fluid flow at the microscale/nanoscale has enabled unprecedented opportunities to develop miniaturized systems for the analysis and manipulation of particles, cells, and biomolecules.

Within this chapter, we introduce three widely used microfluidic approaches for single-cell electrical characterization, namely, dielectrophoresis, electrorotation, and microfluidic impedance cytometry. In dielectrophoresis and electrorotation, electric-field induced mechanical actions (respectively forces and torques) are exploited to manipulate cells. The resulting behavior (respectively displacement and rotation) depends on the geometrical and electrical properties of the cell. Consequently, these cell properties can be estimated by inverse analysis based on suitable modeling approaches. In microfluidic impedance

*Corresponding authors.

https://doi.org/10.1016/B978-0-323-89841-6.00008-6

cytometry, measurements of electrical currents are used in combination with shell-models to characterize cell electrical properties. On the other hand, microfluidic impedance cytometry can use electrical measurements to also characterize nonelectrical particle properties (e.g., size, shape, deformability, velocity and position). Furthermore, characterization may refer to the quantitative estimation of single-cell properties or—a simpler task—to the discrimination between different cell types or subpopulations based on their electrical behavior. Finally, it is worth mentioning that other techniques for single-cell electrical characterization have been developed, which are not covered in this chapter, like for instance conventional patch clamp, nanoprobes, and electrical impedance spectroscopy of trapped cells.[5,6]

The chapter is organized as follows: Section 2 provides an introduction to the dielectric properties of cells and their modeling using Maxwell Mixture Theory, Clausius-Mossotti factor and multi-shell models; Section 3 discusses the basic theory and applications of dielectrophoresis and electrorotation, highlighting their significance for the study of physiological phenomena; Section 4 presents the use of microfluidic impedance cytometry for cell discrimination and characterization, and provides examples of cytometer designs and recent applications; Section 5 concludes the chapter with an overview of future trends.

2 Basic theory

Biological cells are not homogeneous in their composition: typically, cells consist of a nucleus floating in the cytoplasm surrounded by membranes of various compositions. Each domain or layer has its own electrical properties[4,7,8] which can be characterize by two behaviors: resistive and capacitive. Those behaviors are the manifestation of the movements of different charges: free charges (corresponding to the electrical conductivity) for the resistive behavior and bound charges (as represented by the permittivity) for the capacitive behavior which shows the charging mechanism. The mathematical representation of both effects can be visible in the complex permittivity of layer i, ε_i^*, presented in (1):

$$\varepsilon_i^* = \varepsilon_i \varepsilon_0 - j\frac{\sigma_i}{\omega} \tag{1}$$

where ε_i is the relative permittivity, ε_0 the vacuum permittivity (8.854×10^{-12} F/m), σ_i the electrical conductivity (S/m) and $\omega = 2\pi f$ with f the frequency (Hz).

The resulting behavior is frequency dependent: at low frequencies the conductive term is predominant while at higher frequencies the permittivity term is.[9] Modeling a cell requires to take into account the properties of each layer.

2.1 Cell modeling

The equivalent complex permittivity is usually used to model the global behavior of a cell. The most frequently used model for cells is the "single-shell" model in which a cell of radius r_{ext} consists of a cytoplasm surrounded by a membrane of thickness th_{cm}.[9] The corresponding equivalent complex permittivity ε_{eq}^{*} is:

$$\varepsilon_{eq}^{*} = \varepsilon_{cm}^{*} \frac{\left(\frac{r_{ext}}{r_{ext} - th_{cm}}\right)^{3} + 2\left(\frac{\varepsilon_{cp}^{*} - \varepsilon_{cm}^{*}}{\varepsilon_{cp}^{*} + 2\varepsilon_{cm}^{*}}\right)}{\left(\frac{r_{ext}}{r_{ext} - th_{cm}}\right)^{3} - \left(\frac{\varepsilon_{cp}^{*} - \varepsilon_{cm}^{*}}{\varepsilon_{cp}^{*} + 2\varepsilon_{cm}^{*}}\right)} \quad (2)$$

where ε_{cm}^{*} and ε_{cp}^{*} refer, respectively, to the complex permittivities of the cell membrane and cytoplasm.

We pose C_{cm}^{*} the complex capacitance:

$$C_{cm}^{*} = C_{cm} - j\frac{G_{cm}}{\omega} \quad (3)$$

where $C_{cm} = \frac{\varepsilon_{cm}\varepsilon_{0}}{th_{cm}}$ and $G_{cm} = \frac{\sigma_{cm}}{th_{cm}}$ are the specific capacitance (F/m^2) and specific conductance (S/m^2) of the membrane, respectively. As $th_{cm} \ll r_{ext}$, Eq. (2) can be simplified as:

$$\varepsilon_{eq}^{*} = \frac{r_{ext}C_{cm}^{*}\varepsilon_{cp}^{*}}{r_{ext}C_{cm}^{*} + \varepsilon_{cp}^{*}} \quad (4)$$

Depending on the cell of interest, models taking into account more cell components or different cell shapes such as ellipsoid might be needed,[10] as presented in Fig. 1 from Cottet et al.[11]

Cell properties are often not accessible directly and the liquid in which cell are immersed, called suspension or immersion medium, must be taken into account in the modeling.

2.2 Cell suspension and Clausius-Mossotti factor

According to the Maxwell Mixture Theory,[3,12] the effective complex permittivity of a suspension of particles or cells, ε_{mix}^{*}, depends on the volume fraction ϕ, a parametrer representing the volume occupancy by the cell. However, such representation is only valid for a small cell in a homogeneous electric field. ε_{mix}^{*} can be obtained by the Maxwell-Garnett mixing equation if $\phi < 0.1$:

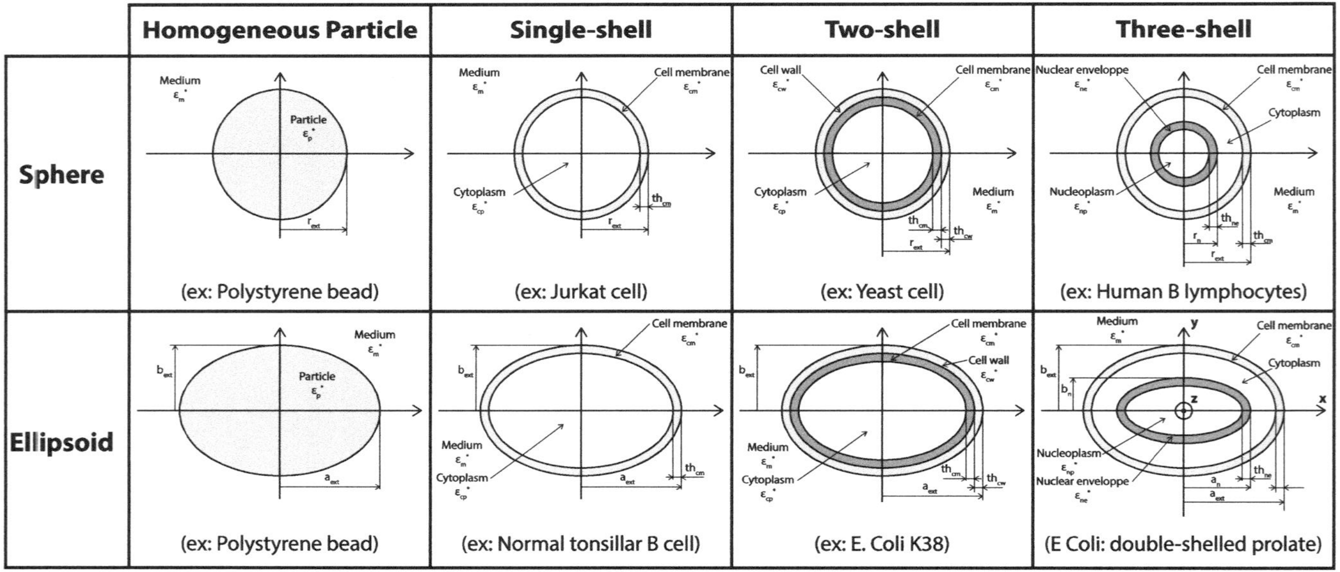

Fig. 1 Different spherical and ellipsoidal cell and particle models (*m*, medium; *p*, particle; *cp*, cytoplasm; *cm*, cell membrane; *cw*, cell wall; *np*, nucleoplasm; *ne*, nuclear envelope). *(Adapted with permission from reference Cottet, J.; Fabregue, O.; Berger, C.; Buret, F.; Renaud, P.; Frénéa-Robin, M. MyDEP: A New Computational Tool for Dielectric Modeling of Particles and Cells. Biophys. J.* ***2019****, 116 (1), 12–18 copyright 2019 Elsevier B.V.)*

$$\frac{\varepsilon_{mix}^* - \varepsilon_m^*}{\varepsilon_{mix}^* + 2\varepsilon_m^*} = \phi \frac{\varepsilon_p^* - \varepsilon_m^*}{\varepsilon_p^* + 2\varepsilon_m^*} \tag{5}$$

where index p and m refer to the particle and the medium, respectively (for a single-shell model $\varepsilon_p^* = \varepsilon_{eq}^*$ from (4)). Alternatively, ε_{mix}^* can be expressed as:

$$\frac{\varepsilon_{mix}^* - \varepsilon_m^*}{\varepsilon_{mix}^* + 2\varepsilon_m^*} = \phi CM(f) \tag{6}$$

where $CM(f)$ is the Clausius-Mossotti factor which reflects the polarizability contrast between the particle and its immersion medium:

$$CM(f) = \frac{\varepsilon_p^* - \varepsilon_m^*}{\varepsilon_p^* + 2\varepsilon_m^*} \tag{7}$$

Eq. (7) can also be rewritten to obtain a direct formulation of ε_{mix}^*, as presented in[12]:

$$\varepsilon_{mix}^* = \varepsilon_m^* \left(1 + 3\phi \frac{CM(f)}{1 - \phi CM(f)}\right) \tag{8}$$

However, for $\phi > 0.1$, the Hanai equation,[12] theoretically valid up to $\phi < 0.8$[13, 14] should be implemented:

$$\left(\frac{\varepsilon_{mix}^* - \varepsilon_p^*}{\varepsilon_m^* - \varepsilon_p^*}\right)\left(\frac{\varepsilon_m^*}{\varepsilon_p^*}\right)^{1/3} = 1 - \phi \tag{9}$$

Unfortunately, there is no direct expression for the calculation ε_{mix}^* from the Hanai formulation for high volume fraction.

The Hanai formulation has proven to be more accurate than the Maxwell-Garnett formulation for higher volume fractions.[15] For more details on the implementation see Refs. 11, 16, 17.

It is important to notice that the calculation of ε_{mix}^* does not depend on the position of particles/cells as the derivation assumes a homogeneous electric field.

Remark: The Clausius-Mossotti factor, also abbreviated as *CM*, plays a pivotal role in the techniques for single-cell electrical characterization. In particular, the dielectrophoretic force is proportional to the real part of *CM* [cf. Eq. (10)], the electrorotation torque is proportional to the imaginary part of *CM* [cf. Eq. (11)], whereas the measured impedance signal depends on the whole complex *CM* [cf. Eq. (18)]. Moreover, these are all proportional to the volume of the particles (either the cube of the radius or the volume fraction), while exhibiting different dependencies on the electric field (gradient of the square of the electric field intensity, square of the electric field intensity or applied voltage, respectively). Fig. 2 represents the real and imaginary parts of *CM* for Human B lymphocytes (parameters taken from Ref. 18) in two media of different electrical conductivities (*A*) $\sigma_m = 0.056$ S/m and (B) $\sigma_m = 1.6$ S/m.

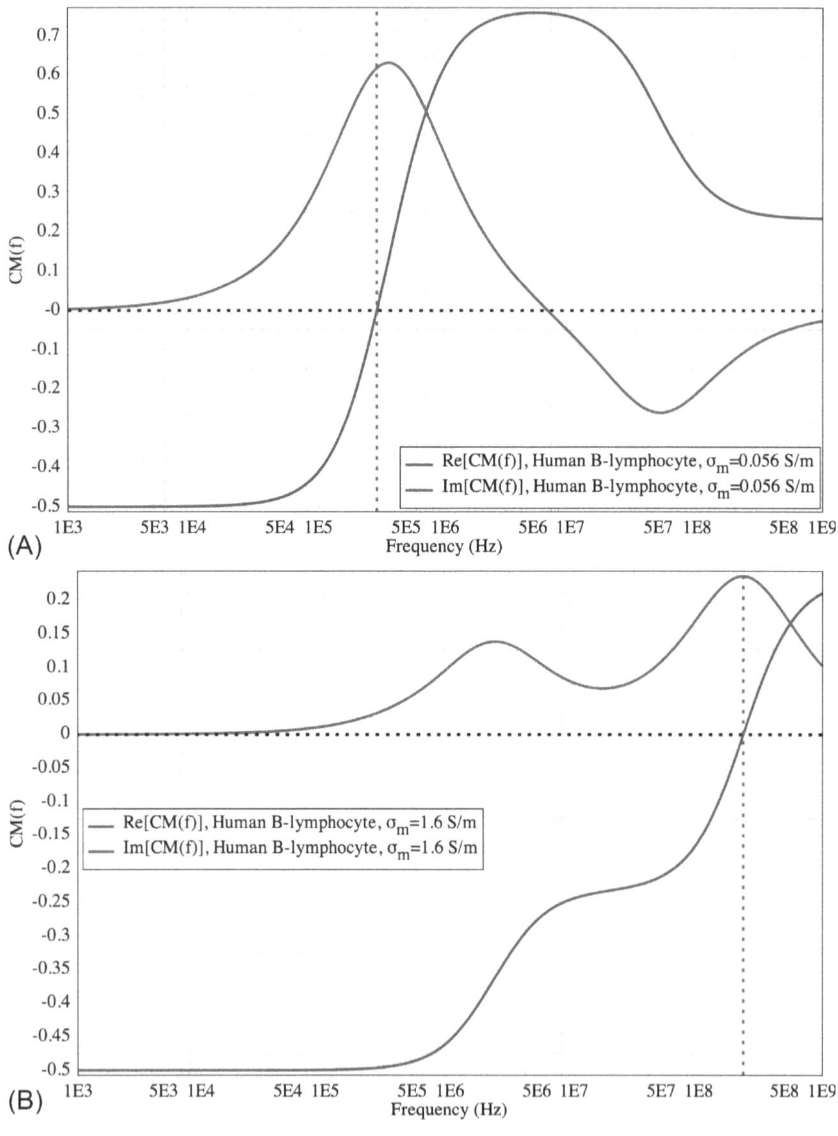

Fig. 2 Real and imaginary parts of the Clausius-Mossotti factor of a Human B-lymphocyte in (A) $\sigma_m = 0.056$ S/m and (B) $\sigma_m = 1.6$ S/m. The vertical red dotted line corresponds to the crossover frequency: (A) 293 kHz and (B) 228 MHz (parameters values can be found in the original work from Ref. 18, graph generated with MyDEP[11]).

Dielectrophoresis, electrorotation, and microfluidic impedance cytometry all exploit the dielectric characteristics of cells and a unified approach to dielectric single-cell analysis is possible.[19] The following sections will present in more details the different techniques and their applications for single-cell characterization.

3 Dielectrophoresis (DEP) and electrorotation (ROT)

3.1 Dielectrophoresis

The term "dielectrophoresis" was first introduced in 1951[20] when Herbert Pohl used it to describe the motion of dielectric particles caused by the interaction with a nonuniform electric field. Different polarization mechanisms operate depending on the properties of the particle and medium as well as the frequency of the electric field. The key mechanism, called the Maxwell-Wagner interfacial polarization effect,[21] is correlated with the formation of an electric double layer at the interface between the particle and the liquid as well as the charge accumulation between layers of different properties.

The resulting time-averaged dielectrophoretic force on a particle, resulting from the interaction between the induced dipole moment **m** and the field gradient, is expressed as:

$$\boldsymbol{F_{DEP}} = 2\pi\varepsilon_m\varepsilon_0 r_{ext}^3 \,\mathrm{Re}\left[CM\left(f\right)\right]\nabla E_{RMS}^2 \tag{10}$$

where E_{RMS} refers to the root mean square value of the electric field **E**. This phenomenon only exists in the presence of a nonuniform electric field because it depends on the gradient of the squared electric field intensity. The force is proportional to the volume of the particle, as well as to the real part of the Clausius-Mossotti factor, $Re[CM(f)]$ which also determines the direction of the force:

- If the particle is more polarizable than its immersion medium ($Re[CM(f)] > 0$), the particle is driven toward the maximum field intensity areas. This corresponds to *positive dielectrophoresis* (pDEP) illustrated in Fig. 3A.
- On the contrary, if the particle is less polarizable than its immersion medium ($Re[CM(f)] < 0$), the particle is moved toward the minimum field intensity areas, which is referred to as *negative dielectrophoresis* (nDEP) illustrated in Fig. 3B.

Additionally, if ($Re[CM(f\text{v})] = 0$), the particle is not affected by the force as it experiences neither pDEP nor nDEP. The specific frequencies at which this occurs are called *crossover frequencies*. The vertical red dotted lines (*grey*

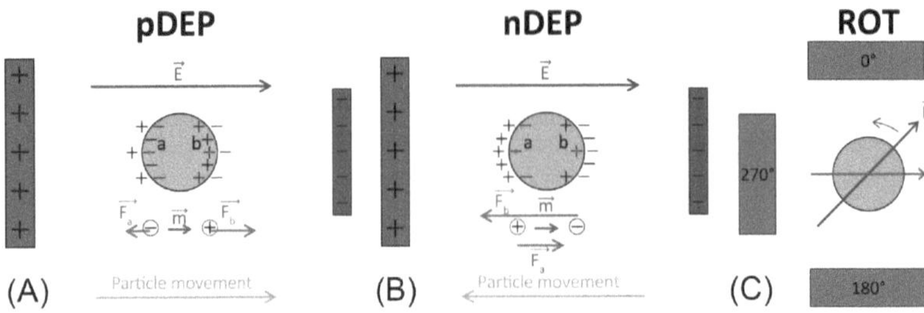

Fig. 3 Principle of (A) pDEP, (B) nDEP, and (C) ROT. **E** represents the electric field, **m** the dipole moment and $\mathbf{F_a}$ and $\mathbf{F_b}$ the Coulomb force ($\mathbf{F}=q\mathbf{E}$ with q the charge) on each barycenter of the charges on each side of the particle.

in print version) in Fig. 2 illustrate the crossover frequencies for Human B lymphocytes for two media of different electrical conductivity: (A) 293 kHz and (B) 228 MHz for $\sigma_m = 0.056$ S/m and $\sigma_m = 1.6$ S/m, respectively.

One of the most popular way to generate DEP in a device is to use patterned electrodes at the bottom of a microchannel. The most common designs are the interdigitated electrodes and castellated interdigitated electrodes. Such electrodes, often deposited on glass, are usually made of gold or platinum with chromium and titanium commonly used as seed layers. Alternatively, ITO (indium tin oxide) is also used because of its transparency. However, DEP generated by direct contact between electrodes and suspending medium can lead to contamination, Joule heating, and electrochemical reactions. To overcome these limitations, different techniques have been developed[22]:

- Insulator-based DEP, abbreviated iDEP, consists in placing insulators between the voltage-applying electrodes. Insulating obstacles will generate nonuniform electric field around them and hence generate DEP forces.
- Another alternative, called contactless DEP or cDEP, keeps the electrodes and the analyte in different channels separated by thin insulating membranes. In this latter method, the side channels are usually filled with a highly conductive liquid. This method is especially suitable for highly conductive samples.
- More recently a method named optical DEP (oDEP) has emerged. Optical images projected onto a photoconductive surface, typically amorphous silicon, create virtual electrodes that can therefore be reconfigured at any time.

3.2 Electrorotation

Electrorotation, also abbreviated ROT, occurs when a polarizable particle is exposed to a rotating electric field. This technique was developed by Arnold and Zimmermann in the 1980s.[23] They proposed to use a four-pole electrode structure where each electrode phase is shifted by 90 degrees between two adjacent electrodes to generate the field as illustrated in Fig. 3C.

When exposed to a rotating electric field, the polarizable particle forms a dipole which should rotate synchronously with the field. However, when the frequency of the electric field if sufficiently high, this synchronism is not allowed due to the long dipole relaxation time. The resulting temporal shift (or phase delay) between the dipole and the field results in a torque exerted on the particle. The resulting time averaged induced torque is as follows:

$$\Gamma_{\mathbf{EL}} = -4\pi r_{ext}^3 \varepsilon_m \varepsilon_0 Im\left[CM\left(f\right)\right]E^2 \mathbf{e_z} \tag{11}$$

where $\mathbf{e_z}$ represents the unit vector normal to the electrode plane.

As the particle rotates, a resistive viscous torque caused by the surrounding fluid is experienced by the particle[24]:

$$\Gamma_\eta = -8\pi\eta r_{ext}^3 \Omega_0\left(f\right)\mathbf{e_z} \tag{12}$$

where η is the dynamic viscosity of the medium and Ω_0 is the constant angular velocity of the particle.

When the induced torque and resistive viscous torque are in equilibrium, the rotation rate for a spherical particle is given by:

$$\Omega_0\left(f\right) = -\frac{\varepsilon_0\varepsilon_m}{2\eta} Im\left[CM\left(f\right)\right]E^2 \tag{13}$$

The particle rotates against the field direction for $Im[CM(f)] > 0$ due to the minus sign. Dielectric parameters of a cell can therefore be obtained by curve fitting procedures of the electrorotation spectrum.

3.3 Applications for characterization

DEP and ROT have been extensively used for dielectric characterization of cells.

DEP can be used as a characterization tool for three different techniques[25]:

- Crossover frequency measurement. Crossover frequencies are dependent of the dielectric properties of the cells and the medium.[26,27] For mammalian cells, two crossover frequencies are generally observed.[22] The first crossover frequency (nDEP to pDEP), abbreviated fc1, is determined mostly by cell shape, size, and membrane properties and is typically observed in the 10–100 kHz range.[28] The second crossover frequency (pDEP to nDEP), abbreviated fc2, is mainly determined by internal cell parameters, such as cytoplasmic conductivity and permittivity and is typically above 10 MHz in low conductivity media.[29]

The extraction of the dielectric properties of the cell of interest is usually based on the exploitation of datasets comprising various media conductivities and thus different crossover frequencies. Subsequent fitting procedure with the appropriate shell model allows the determination of the different dielectric parameters. The technique enables to characterize the following parameters: C_{cm}, G_{cm}, σ_{cm}, ε_{cm}, ε_{cp}, σ_{cp} (and even the dielectric parameters ε_{cw} and σ_{cw} of the cell wall when a two-shell model is used).

Many examples can be found in the literature[30–40]

- Dielectrophoretic capture-voltage spectrum. This method uses a balance between drag force and DEP force to characterize cells.[41, 42] In this technique, a low conductive media is usually used to enhance pDEP. Contrary to the crossover frequency method, the electric field frequency is fixed and the voltage is varied until trapped cells are observed on the electrode. After repeating the procedure for all the frequencies of interest, the *CM* factor is obtained and can be fitted to obtain the cell properties. The parameters obtained with this method are σ_{cm}, ε_{cm}, ε_{cp}, and σ_{cp}.
- Dielectrophoretic collection spectra. In this technique developed by Gascoyne et al.[43] the rate at which cells accumulate in an interdigitated electrode configuration while experiencing positive and negative dielectrophoresis is tracked and recorded. The collection rate and frequency data are then plotted and fitted with the corresponding shell model to extract the dielectric properties.[44–46] The parameters obtained with this method are C_{cm}, σ_{cm}, and ε_{cm}.

Electrorotation spectra, which are normalized rotation rate as a function of frequency, are often used in combination with a single-shell model and fitting algorithms to extract dielectric properties of cells. The parameters usually extracted with this method are C_{cm}, ε_{cp}, and σ_{cp}.[47–51] See the recent review for more information on the different ROT methods used for the extraction of dielectric parameters.[52]

It is important to mention that DEP and ROT are often combined on the same chip with the same electrodes because DEP can used to stop and center cells before performing ROT.[49, 53, 54]

3.4 Signification of the differences

After obtaining the dielectric parameters through one of the previously mentioned methods, the differences between cells were analyzed to understand how they correlate with the physiological phenomena.

The most typical characterization performed is viability.Viable and dead cells have been reported to exhibit different electrical properties. In particular viable cells normally have a low conductive membrane while the interior has a high electrical conductivity.[55] For dead cells, the membrane becomes permeable and its electrical conductivity increases while the electrical conductivity of the interior of the cell decreases.[56, 57] Additionally, the conductivity of the cell nucleus is also reduced.[53] More details can be found in this recent review.[58]

Specific membrane capacitance can also reveal differences in cell surface morphology.[59, 60] A decrease in specific membrane capacitance may reflect a smoothing of the membrane, for example, through the loss of microvilli.[61] Specific membrane capacitance has been shown to increase when cells undergo the different stages of cancer.[36]

Cytoplasmic conductivity correlates with physiologic phenomena such biological rhythms, drug sensitivity, and resistance,[45, 62] intracellular lipid content for algae,[63] antimicrobial resistance,[51] and mitochondrial morphology.[64]

Some processes, such as apoptosis, have been shown to be detectable and associated with a decrease in specific membrane capacitance and in cytoplasmic conductivity.[31, 65]

4 Microfluidic impedance cytometry

Due to their simplicity and label-free nature, electrical sensing techniques are extremely attractive for the analysis and characterization of particles and cells. Since the introduction of the Coulter principle for particle counting,[1] electrical sensing has been exploited in several applications in life sciences, medicine, and environmental monitoring. In particular, in the past two decades, the development of microfabrication technologies has enabled the implementation of microfluidic impedance cytometers for the high-throughput label-free electrical characterization of blood cells, mammalian cell lines, microorganisms, and plant cells.[66–70]

4.1 Discrimination of cell subpopulations

To explain the basic principles of microfluidic impedance cytometry and its use as a label-free tool for cell phenotyping, an exemplary case study is discussed herein. The example is adapted from De Ninno et al.[71] and is relevant to the characterization of viable, necrotic, and apoptotic cell subpopulations. Canonical hallmarks of apoptotic cell death include cell shrinkage, maintenance of plasma membrane integrity, nuclear fragmentation, and formation of apoptotic bodies. In contrast, necrosis is characterized, among other features, by an early swelling and increased permeability with breakdown of the plasma membrane integrity. Accordingly, different sizes and dielectric properties are expected for apoptotic, necrotic, and viable cell subpopulations, and this should reflect in different impedance signatures.

Let us consider the system in Fig. 4A: cells suspended in a conductive buffer flow one-by-one through a microfluidic channel, where an electric field is established by using three coplanar electrodes integrated into the channel floor. In particular, an AC multi-frequency voltage signal is applied to the central electrode and the differential current collected from the adjacent electrodes is acquired. The latter current is conditioned by a transimpedance amplifier and lock-in demodulated, so that for each frequency channel a complex signal is obtained, composed by a real (in-phase) and an imaginary (out-of-phase) component. Depending on the frequency, different cell properties are probed. While the electrical double layer around the electrodes screens the field at frequencies below 100 kHz, the signal is chiefly determined by the cell size in the 0.1–1 MHz range due to complete membrane-induced screening. Polarization of the plasma membrane in the 2–10 MHz range offers information on the membrane morphology based on measured capacitance. In the 10–30 MHz range, the membrane is minimally polarized, so that the cytoplasmic conductivity and permittivity provide information on the organelles, while at higher frequencies, the response is dependent on nucleus properties.[68, 72]

In the example reported here, a low frequency (0.5 MHz) and a high frequency (10 MHz) channel are considered (Fig. 4B). Wandering baseline and noise can be present in those data streams, so that detrending and denoising are typically performed. Due to the differential measurement scheme, the passage of each flowing cell is recorded as a pair of opposite pulses, in each frequency channel (Fig. 4C). Different segmentation approaches have been used to identify cell passages in the data stream, ranging from threshold-based peak finding to tailored correlation-based approaches. For each detected event, a feature extraction is performed. Although the peak amplitude varies significantly among the different frequency channels and between the real and imaginary signal parts, the peak-width and the peak-to-peak time of the measured signals are quite similar. Accordingly, a unique bipolar Gaussian template with frequency-dependent amplitude (Fig. 4D) can be used to process all frequency channels at the same time[71]:

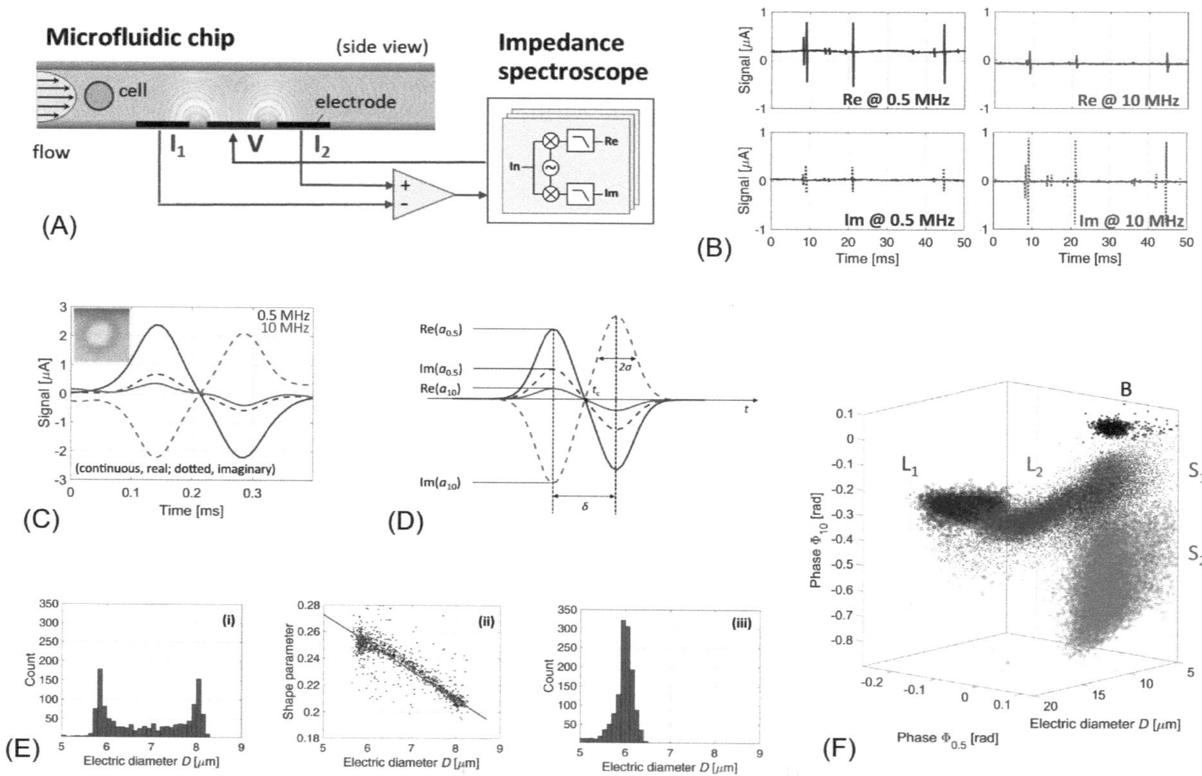

See the figure legend on next page.

Fig. 4 Example of label-free subpopulation analysis by means of microfluidic impedance cytometry (see text for details). (A) Schematic representation of a microfluidic impedance cytometer with three coplanar electrodes operated in differential measuring mode. (B) Data stream portions (real and imaginary parts of the current at 0.5MHz, in *blue* (*light grey in print version*), and 10MHz, in *red* (*grey in print version*), respectively). (C) Example of cell event signal, along with a snapshot of the flowing cell (0.5 and 10MHz frequency channels are shown in *blue* and *red*, respectively; real and imaginary parts of the current are shown as continuous and dotted lines, respectively). (D) Bipolar Gaussian template with complex frequency-dependent amplitude used to fit the cell-event signals. (E) Correction for positional dependence: (i) histogram of 6-μm bead electric diameter, (ii) bidimensional histogram of shape parameter against electric diameter, (iii) histogram of 6-μm bead electric diameter after correction. (F) Scatter plot of the electric diameter, electric phase at 0.5MHz and electric phase at 10MHz, showing the presence of five subpopulations (see text for details). *(Panels A, C, and F are adapted with permission from De Ninno, A.; Reale, R.; Giovinazzo, A.; Bertani, F. R.; Businaro, L.; Bisegna, P.; Matteucci, C.; Caselli, F. High-Throughput Label-Free Characterization of Viable, Necrotic and Apoptotic Human Lymphoma Cells in a Coplanar-Electrode Microfluidic Impedance Chip. Biosens. Bioelectron. 2020, 150, 111887, copyright 2020 Elsevier B.V.)*

$$s(t)_f = a_f\left(e^{-\frac{(t-(t_c-\delta/2))^2}{2\sigma^2}} - e^{-\frac{(t-(t_c+\delta/2))^2}{2\sigma^2}}\right) \tag{14}$$

The template is characterized by the complex frequency-dependent amplitude a_f, the peak-width control σ, the peak-to-peak time δ, and the central time t_c. The peak-to-peak time δ can be used to compute cell velocity, by using geometric information (i.e., electrode pitch). The complex amplitude a_f is typically converted in modulus and phase. Since at low frequencies signal amplitude is proportional to cell volume, its cube root provides an electrical measure of the cell diameter (the electric diameter). These features (electric diameter and phases at different frequencies) can be used to discriminate between subpopulations. However, their accuracy is limited by position blurring, that is, identical cell flowing along different trajectories provide different signals. This is a consequence of the nonuniformity of the electric field, which is quite relevant in the present coplanar electrode layout. As an example, the measured electrical diameter of 6-μm beads is reported in Fig. 4 [E, panel (i)], and exhibits significant spread and skewness. Several solutions are possible to mitigate position blurring, including the use of electrical metrics less sensitive to particle position (e.g., electrical opacity), the implementation of active or passive particle focusing mechanisms (e.g., DEP-focusing, acoustophoretic focusing, or inertial focusing), optimization of channel geometry, or the use of compensation strategies based on electrical metrics for particle position. An extensive discussion on this topic can be found in the recent review by Daguerre et al.[73] In the example discussed herein, we consider the use of a compensation strategy based on the signal shape-parameter σ/δ.[74] Indeed, the latter is a metric that correlates with particle vertical position, and hence can be used to purge out position blurring from the measured electric features by means of simple calibration [Fig. 4E, panels (ii) and (iii)]. After compensation, the scatter plot of the electrical features (electric diameter, phase at 0.5 MHz and phase at 10 MHz) enables the identification of five subpopulations (Fig. 4F). As motivated in the original work[71] by means of comparison with flow cytometry results, they reasonably correspond to reference beads (B), viable cells (L_1), necrotic cells (L_2), and two subpopulations of apoptotic bodies/cell debris (S_1, S_2). Viable and necrotic cells exhibit a higher electric diameter with respect to apoptotic bodies/cell debris and can be distinguished from each other by their low-frequency phase. The two subpopulations of apoptotic bodies/cell debris can be discriminated by means of their high-frequency phase.

4.2 Impedance spectra fitting

In addition to discrimination of cell subpopulations, impedance cytometry can be used to characterize particle dielectric properties by fitting impedance spectra. Let us consider a cytometer design with two pairs of facing electrodes located at the top and the bottom of the microchannel. The voltage V_{appl} is applied to the top electrodes and the differential current from the bottom electrodes I_{diff} is measured (like in Fig. 5A). When the cell is centered between the first electrode pair, neglecting stray capacitances, such differential current can be written as:

$$I_{\text{diff}} = \frac{V_{\text{appl}}}{2Z_{\text{dl}} + Z_{\text{med}}} - \frac{V_{\text{appl}}}{2Z_{\text{dl}} + Z_{\text{mix}}} \approx V_{\text{appl}} \frac{\Delta Z}{\left(2Z_{\text{dl}} + Z_{\text{med}}\right)^2} \tag{15}$$

where Z_{dl} is the electrode double layer impedance, Z_{med} is the medium impedance, Z_{mix} is the impedance of the mixture composed by the cell and its surrounding medium, and $\Delta Z = (Z_{mix} - Z_{med})$ is assumed $\ll Z_{med}$. The impedances Z_{med} and Z_{mix} are usually expressed as:

$$Z_{\text{med}} = \frac{1}{j\omega\varepsilon^*_{med} G} \tag{16}$$

and

$$Z_{\text{mix}} = \frac{1}{j\omega\varepsilon^*_{mix} G} \tag{17}$$

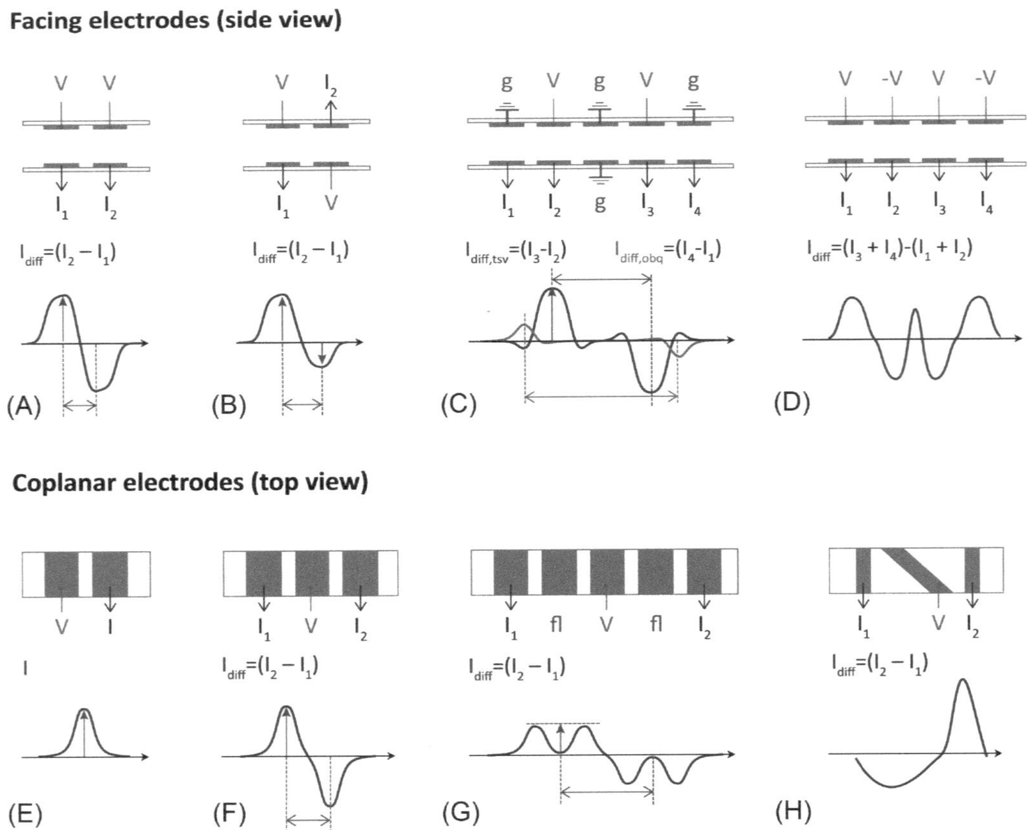

Fig. 5 Examples of chip designs and wiring schemes, based on (A–D) facing electrodes or (E–H) coplanar electrodes (*V*, applied voltage; *I*, measured current; *g*, grounded electrode; *fl*, floating electrode; *diff*, differential current; *tsv*, transverse current path; *obq*, oblique current path). Idealized current signals obtained via finite element simulations are also shown, with an indication of some of the extracted features (e.g., amplitude(s) and transit time(s)). See text and original works for details.

where G is a geometric factor accounting for electrode dimensions, channel height and fringing fields. By using Maxwell Mixture Theory to express ε^*_{mix}, under the assumption of small volume fraction ϕ, the measured signal S can be approximated with (cf. e.g., Ref. 68):

$$S = HI_{\text{diff}} \approx -H\frac{V_{\text{appl}}}{Z_{\text{med}}}\frac{1}{\left[1+\left(2Z_{\text{dl}}/Z_{\text{med}}\right)\right]^2}3\phi CM \tag{18}$$

where H denotes the frequency response of the acquisition system and CM is the Clausius-Mossotti factor.

Eq. (18) shows that the measured signal S approximately linearly depends on ϕ, conveying information on cell size, and CM, conveying information on cell dielectric properties. However, S also depends on details of the measurement system (e.g., transfer function, applied voltage, double-layer impedance, and medium properties). The task of extracting ϕ and CM from the measured signal S is usually accomplished by using reference beads. Denoting by S^{cell} and S^{bead} the signals respectively recorded under cell or reference-bead passage, Eq. (18) yields:

$$\frac{S^{\text{cell}}}{S^{\text{bead}}} = \frac{CM^{\text{cell}}\phi^{\text{cell}}}{CM^{\text{bead}}\phi^{\text{bead}}} \tag{19}$$

where details of the measurement system have been effectively cancelled out. Assuming that the cell is spherical and that:

$$\frac{\phi^{\text{cell}}}{\phi^{\text{bead}}} = \left(\frac{r^{\text{cell}}}{r^{\text{bead}}}\right)^3 \tag{20}$$

where r^{cell} and r^{bead}, respectively, denote cell and bead radii,[75] Eq. (19) is recast as:

$$CM^{\text{cell}}\left(\frac{r^{\text{cell}}}{r^{\text{bead}}}\right)^3 = CM^{\text{bead}}\frac{S^{\text{cell}}}{S^{\text{bead}}} \tag{21}$$

Here CM^{bead} and r^{bead}, relevant to reference beads, can be regarded as known quantities, whereas CM^{cell} and r^{cell} are to be determined. The normalized data in Eq. (21) can be fitted using shell models, to characterize the cell dielectric properties from impedance spectra. Both population-averaged[76] and single-cell approaches[75, 77] have been presented in the literature. Note that the small volume fraction limit means that this method only works for systems with a high signal-to-noise ratio where the channel dimensions are much larger than the cell.[75] Finally, it is worth mentioning that modeling approaches based on equivalent circuits have been also used to characterize the dielectric properties of the flowing cells.[78]

4.3 Designs of impedance cytometers

A number of cytometer designs have been proposed in the literature. Different designs provide different signal shapes, which carry different information content. Fig. 5 shows a gallery of possible designs, based on straight rectangular channels with either facing electrodes (panels A–D) or coplanar electrodes (E–H). For each case, a schematic of the chip layout and of the wiring scheme is presented, along with an idealized measured current signal obtained by finite element simulations. In general, facing electrode layouts have higher sensitivity and suffer less from positional dependence, while coplanar electrode designs are easier to fabricate. Panel (A) shows the most typical design with two pairs of facing electrodes, in which voltage is applied to the top electrodes and the differential current from the bottom electrodes is measured.[72] The latter current has a bipolar Gaussian shape with positive and negative pulses of equal amplitude. Such (complex frequency-dependent) amplitude can be used for particle characterization, while the peak-to-peak time yields particle velocity. With a straightforward modification of the wiring scheme (i.e., applying voltage to diagonally opposite electrodes, as in panel (B)), the resulting signal has a bipolar Gaussian shape with unequal positive and negative pulses.[79] Interestingly, the pulse amplitude relative difference correlates with particle vertical position and can be used to mitigate position blurring or to implement an electrical position sensor. Another way to obtain an electrical metric that correlates with particle vertical position is shown in panel (C): five pairs of facing electrodes are used to measure two differential currents with oblique or transverse orientation. By geometric arguments, the ratio between their respective transit times is close to one for particles traveling near the top electrodes, and increases with particle distance from the top of the channel.[80] The configuration in panel (D) has four pairs of electrodes: two of the top electrodes are driven with an AC voltage, whilst the other two top electrodes are driven with the same voltage signal, but phase shifted by 180 degrees. Thus the measured currents $(I_3 + I_4)$ and $(I_1 + I_2)$ are zero when there are no particles in the channel. This allows higher gain in the acquisition system, enabling smaller particles to be measured in a large channel.[81]

Coplanar electrodes are used in both absolute and differential measurement schemes. In the former case, two electrodes are used: voltage is applied to one electrode and the current flowing through the other electrode is collected (panel (E)). The passage of a flowing particle is recorded as a current pulse in each frequency channel. Pulse magnitude and phase are extracted for particle characterization. Absolute measurement schemes are widely used in applications seeking simplicity and portability. The standard differential measurement scheme uses three coplanar electrodes (panel (F)) and has been discussed in detail in Section 4.1. By introducing a floating electrode between the stimulating and sensing electrodes, as in panel (G), the resulting signal exhibits a bipolar double-Gaussian shape. The relative prominence of the peaks with respect to the saddle in between them correlates with particle vertical position: it is more pronounced for particles traveling near the electrodes and less pronounced for particles far from the electrodes.[82] Information on particle lateral position can be obtained by using N-shaped electrodes[83] (panel (H)), on the basis of simple geometric relationships.

The gallery presented in Fig. 5 is by no means exhaustive. Other design solutions are possible, for example, liquid electrodes (electrodes positioned at the bottom of dead-end chambers placed on each side of the main channel),[84] liquid electrodes in a cross-shape,[85] Coulter-networks,[86] 3D free-standing electrodes.[87] It is also possible to modulate channel width, to achieve constriction, relaxation, or extension zones,[88, 89] to perform impedance-based single-cell mechanical phenotyping,[90, 91] typically on the basis of transit times and transit-time ratios. Microtubular devices with integrated electrodes are also being developed.[92] For an in-depth presentation of recent microfluidic cytometer designs, the reader is referred to Honrado et al.[68]

4.4 Applications

Several applications of microfluidic impedance cytometry have been reported in the literature, including but not limited to[68]: animal and plant-cell viability assays; cell-cycle monitoring; characterization of activation, differentiation, and diabetes response of immune cells; monitoring stem-cell differentiation; characterization of parasite-infected red blood cells; testing drug sensitivity and tumorigenicity of cancer cells; detection and antibiotic susceptibility of bacteria; yeasts cell morphological analysis; impedance-based deformability cytometry; impedance-based position detection; droplet characterization; and classification of pollen grains. In this section, four application examples from the recent literature are described (Fig. 6).

The first example (Fig. 6A) shows the application of an impedance-based position sensor to monitor particle inertial focusing, in a completely

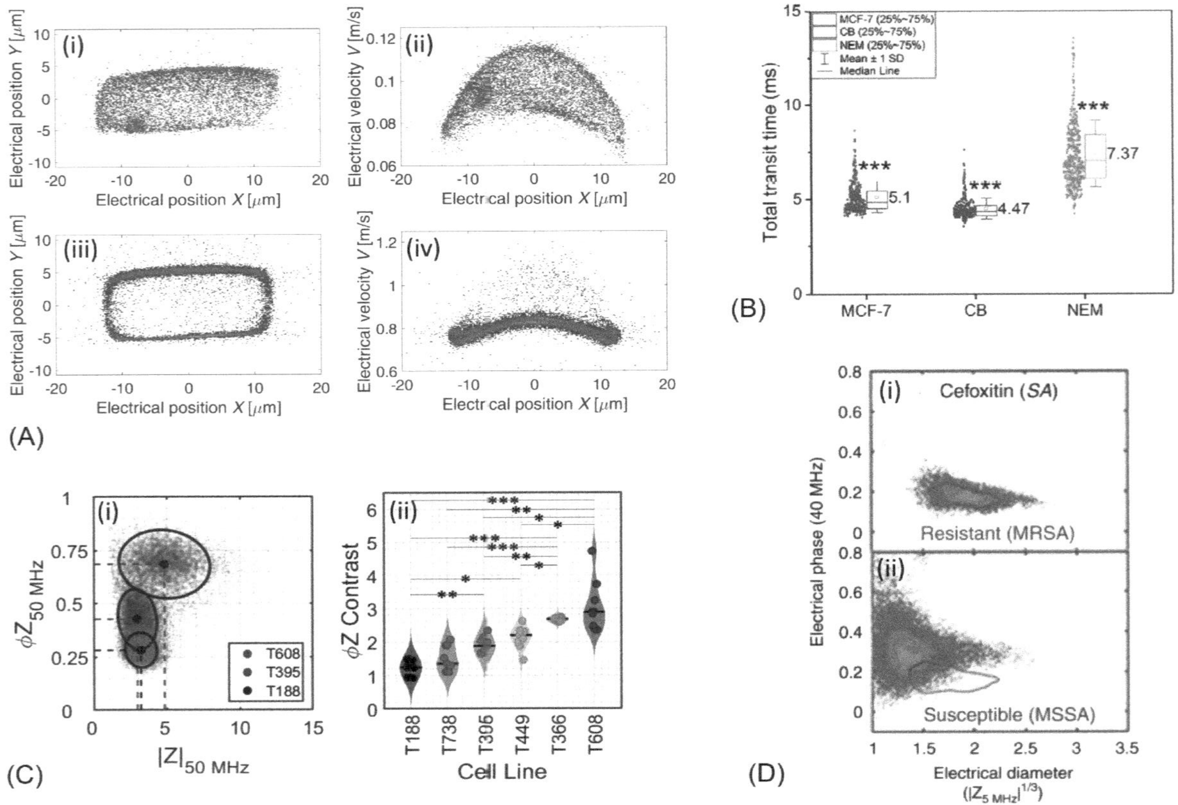

See the figure legend on opposite page.

optics-fee way.[93] Electrical estimates (i.e., estimates obtained by processing impedance traces) of the lateral position (X), vertical position (Y), and velocity (V) of single flowing beads are obtained by using an innovative microfluidic cytometer design (based on a combination of the concepts in Fig. 5 panels (C) and (G)). When the particle Reynold's number is low (0.31), the beads are randomly distributed in the channel cross-section and the full parabolic velocity profile typical of laminar flow is recognized (Fig. 6A subpanels (i) and (ii)). On the other hand, at higher particle Reynold's number (3.14), the beads are focused on an annular region close to the channel walls and the velocity profile changes accordingly (Fig. 6A subpanels (iii) and (iv)).

The second example deals with the biophysical phenotyping of single cells by means of a differential multiconstriction microfluidic device with 3D electrodes.[91] The approach enables the simultaneous characterization

---

Fig. 6 Examples of microfluidic impedance cytometry applications. (A) Electrical position detection of particles for inertial focusing studies: (i) density plot of electrical position Y vs. electrical position X and (ii) density plot of electrical velocity V vs. electrical position X, at low (i.e., 0.31) particle Reynold's number; (iii) and (iv), density plots of the same quantities at high (i.e., 3.14) particle Reynold's number. (B) Impedance-based transit time as a metric for cell deformability: total transit time of cells passing through four successive constrictions for three cell populations (MCF-7, CB, NEM); the numbers near the box charts indicate the mean values of total transit time; *** means there is a significant difference among cell populations ($P < .001$), based on the Mann-Whitney test. (C) Impedance phase contrast for stratification of pancreatic cells tumorigenicity: (i) measurements of impedance magnitude vs. phase at 50 MHz for various pancreatic ductal adenocarcinoma cell lines; (ii) comparison of impedance phase contrast between cell lines, where significant differences were observed (*$P < .003$, **$P < .0007$, ***$P < .00007$). (D) Fast impedance-based antimicrobial susceptibility test[81]: density plot of electrical phase at 40 MHz vs. electrical diameter at 5 MHz, for (i) resistant and (ii) susceptible *Staphylococcus aureus* after exposure to Cefoxitin antibiotic; the original unexposed contour is also shown. *((A) Adapted from Reale, R.; De Ninno, A.; Businaro, L.; Bisegna, P.; Caselli, F. Electrical Measurement of Cross-Sectional Position of Particles Flowing Through a Microchannel. Microfluid. Nanofluid.* ***2018****, 22 (4). (B) Adapted with permission from Yang, D.; Zhou, Y.; Zhou, Y.; Han, J.; Ai, Y. Biophysical Phenotyping of Single Cells Using a Differential Multiconstriction Microfluidic Device With Self-Aligned 3D Electrodes. Biosens. Bioelectron.* ***2019****, 133, 16–23, copyright 2019 Elsevier B.V. (C) Adapted with permission from McGrath, J. S.; Honrado, C.; Moore, J. H.; Adair, S. J.; Varhue, W. B.; Salahi, A.; Farmehini, V.; Goudreau, B. J.; Nagdas, S.; Blais, E. M.; Bauer, T. W.; Swami, N. S. Electrophysiology-Based Stratification of Pancreatic Tumorigenicity by Label-Free Single-Cell Impedance Cytometry. Anal. Chim. Acta* ***2020****, 1101, 90–98, copyright 2019 Elsevier B.V. (D) Adapted from D.C., Spencer; T.F., Paton; K.T., Mulroney; et al. A Fast Impedance-Based Antimicrobial Susceptibility Test. Nat. Commun.* ***2020****, 11 (1), 5328.)*

of the deformability, electrical impedance, and relaxation index of single cells. Fig. 6B reports the total transit time of cells passing through four successive constrictions for three populations of MCF-7 breast cancer cells: untreated (MCF-7), cytochalasin B treated (CB) and *N*-ethylmaleimide treated (NEM). It can be inferred that the NEM-MCF-7 cells are the least deformable ones as these cells require the longest time to pass through constriction regions. On the contrary, CB-treated MCF-7 cells are the most deformable ones. The deformability of normal MCF-7 cells sits in between NEM-treated and CB-treated MCF-7 cells.

The third example[94] demonstrates the phenotypical stratification of patient-derived, pancreatic tumor xenografts based on their tumorigenicity, by using a standard cytometer design with two facing electrode pairs (as in Fig. 5A). In particular, single cells derived from primary pancreatic tumors versus those from liver metastasis were analyzed with a novel impedance phase contrast metric, which is based on impedance phase variations in the high and low frequency range. Variations in this metric were found to be related to electrophysiology of the cell interior and to be systematically altered as a function of tumorigenicity (Fig. 6C).

The last example describes a label-free antimicrobial susceptibility test that can deliver results within an hour.[81] The bacteria are incubated in the presence of an antibiotic for 30 min, and then approximately 10^5 cells are analyzed one-by-one with microfluidic impedance cytometry for 2–3 min (using the configuration in Fig. 5D). The measured electrical characteristics reflect the phenotypic response of the bacteria to the mode of action of a particular antibiotic, in a 30-min incubation window. As an example, Fig. 6D shows the scatter plots of electrical phase against electrical diameter for resistant and susceptible *Staphylococcus aureus* after exposure to Cefoxitin antibiotics.

5 Recent trends and future developments

Nowadays, biosensors based on DEP, ROT, or microfluidic impedance cytometry have gained an important role for single-cell electrical characterization. As we have seen in this chapter, several successful applications, dealing with different cell types, have been reported in the scientific literature. Besides that, commercial products are now available on the market, testifying the readiness of these technologies. CELLSEARCH (Menarini) Inc., Precision For Medicine LLC. (formerly known as ApoCell using ApoStreamTM CTC enrichment technology), and ClearCell of Biolidics

Limited (formerly known as Clearbridge BioMedics) have commercialized their DEP-based technologies for the separation of CTCs. Berkeley Lights Inc. has successfully commercialized its oDEP products for cell separation. Commercial microfluidic impedance cytometers are also available. Amphasys A.G. (Switzerland) has commercialized their product line AmphaZ, especially for pollen analysis and viability assays, SBT Instruments (Denmark) has developed BactoBox for bacteria detection, and WeareCELLIX (Ireland) is going to launch the Inish Cell Analyser.

On the other hand, the research field is quite lively, with research efforts being directed toward further enhancement of the techniques. Efforts to improve throughputs and recovery rates as well as to limit the exposure time to modified media (typically with low electrical conductivities for DEP analysis) will allow DEP and ROT-based technologies to go toward more commercial products. Automation and parallelization will also play key roles in achieving these goals. In microfluidic impedance cytometry, the extension of the frequency rate beyond several MHz and the possibility to perform spectroscopy at an increased number of simultaneous frequencies could enable a more comprehensive characterization of subcellular features. The development of designs suitable for highly size-heterogeneous samples and designs providing a multiparametric characterization embracing cell dielectric properties, motion-related properties, shape, orientation, and deformability is also attractive to improve the discrimination of different cell subpopulations.

The development of simplified designs and alternative fabrication approaches, possibly cleanroom free, enabling low-cost disposable devices is also highly desirable. This would promote a widespread use of the technologies and the development of point-of-care solutions. On the other hand, integrated multifunctional platforms combining DEP, ROT, and microfluidic impedance cytometry among them or with other techniques on chip would be an opportunity toward the development of a microfluidic chip that encompasses the entire single-cell workflow from start to finish (i.e., cell enrichment/selection, characterization, readout).[95]

Finally, future developments are expected from the synergistic convergence of microfluidics and machine learning.[96–98] Steps in these directions have already been taken. For instance, machine learning has been used in DEP for sorting decisions[99] and in ROT to improve the extraction of cell dielectric features.[100] In microfluidic impedance cytometry, machine learning approaches have been used either to perform population clustering based on electrical metrics[91, 101] or toward real-time analysis of raw electrical signals,[102] which could enable online sorting of target subpopulations in heterogeneous samples. Overall, microfluidics, electric fields and machine learning are expected to play a pivotal role in the coming years, to bring single-cell analysis to its next level.

Acknowledgment

J.C. acknowledges support from the Office of Naval Research (N00014-18-1-2632). F.C. wishes to thank Paolo Bisegna for fruitful discussion.

References

1. Coulter, W. H. *Means for Counting Particles Suspended in a Fluid, Coulter, W. H*; 1953. US11281949A.
2. Foster, K. R.; Schwan, H. P. Dielectric Properties of Tissues and Biological Materials: A Critical Review. *Crit. Rev. Biomed. Eng.* **1989**, *17* (1), 25–104.
3. Maxwell, J. C. *A Treatise on Electricity and Magnetism*; Cambridge University Press: Cambridge, 1873.
4. Pauly, H.; Schwan, H. P. Über die Impedanz einer Suspension von kugelförmigen Teilchen mit einer Schale. *Z. Naturforsh. B* **1959**, *14* (2), 125–131.
5. Mansor, M. A.; Ahmad, M. R. Single Cell Electrical Characterization Techniques. *Int. J. Mol. Sci.* **2015**, *16* (6), 12686–12712.
6. Feng, Y.; Huang, L.; Zhao, P.; Liang, F.; Wang, W. A Microfluidic Device Integrating Impedance Flow Cytometry and Electric Impedance Spectroscopy for High-Efficiency Single-Cell Electrical Property Measurement. *Anal. Chem.* **2019**, *91* (23), 15204–15212.
7. Fricke, H. A Mathematical Treatment of the Electric Conductivity and Capacity of Disperse Systems ii. The Capacity of a Suspension of Conducting Spheroids Surrounded by a Non-Conducting Membrane for a Current of Low Frequency. *Phys. Rev.* **1925**, *26* (5), 678–681.
8. Fricke, H. The Complex Conductivity of a Suspension of Stratified Particles of Spherical or Cylindrical Form. *J. Phys. Chem.* **1955**, *59* (2), 168–170.
9. Gagnon, Z. R. Cellular Dielectrophoresis: Applications to the Characterization, Manipulation, Separation and Patterning of Cells. *Electrophoresis* **2011**, *32* (18), 2466–2487.
10. Kakutani, T.; Shibatani, S.; Sugai, M. Electrorotation of Non-Spherical Cells: Theory for Ellipsoidal Cells With an Arbitrary Number of Shells. *Bioelectrochem. Bioenerg.* **1993**, *31* (2), 131–145.
11. Cottet, J.; Fabregue, O.; Berger, C.; Buret, F.; Renaud, P.; Frénéa-Robin, M. MyDEP: A New Computational Tool for Dielectric Modeling of Particles and Cells. *Biophys. J.* **2019**, *116* (1), 12–18.
12. Markel, V. A. Introduction to the Maxwell Garnett Approximation: Tutorial. *J. Opt. Soc. Am. A Opt. Image Sci. Vis.* **2016**, *33* (7), 1244–1256.
13. Hanai, T. Electrical Properties of Emulsions. In *Emulsion Science*; Sherman, P., Ed.; Academic Press: London, 1968; pp. 353–478.
14. Hanai, T.; Imakita, T.; Koizumi, N. Analysis of Dielectric Relaxations of w/o Emulsions in the Light of Theories of Interfacial Polarization. *Colloid Polym. Sci.* **1982**, *260* (11), 1029–1034.

15. Mansoorifar, A.; Ghosh, A.; Sabuncu, A. C.; Beskok, A. Accuracy of the Maxwell–Wagner and the Bruggeman–Hanai Mixture Models for Single Cell Dielectric Spectroscopy. *IET Nanobiotechnol.* **2017**, *11* (7), 874–882.
16. Hanai, T.; Asami, K.; Koizumi, N. Dielectric Theory of Concentrated Suspension of Shell-Spheres in Particular Reference to the Analysis of Biological Cell Suspensions. *Bull. Inst. Chem. Res., Kyoto Univ.* **1979**, *57* (4), 297–305.
17. Irimajiri, A.; Suzaki, T.; Asami, K.; Hanai, T. Dielectric Modeling of Biological Cells. Models and Algorithm. *Bull. Inst. Chem. Res., Kyoto Univ.* **1991**, *69*, 421–438.
18. Yang, J.; Huang, Y.; Wang, X. J.; Wang, X. B.; Becker, F. F.; Gascoyne, P. R. C. Dielectric Properties of Human Leukocyte Subpopulations Determined by Electrorotation as a Cell Separation Criterion. *Biophys. J.* **1999**, *76* (6), 3307–3314.
19. Valero, A.; Braschler, T.; Renaud, P. A Unified Approach to Dielectric Single Cell Analysis: Impedance and Dielectrophoretic Force Spectroscopy. *Lab Chip* **2010**, *10* (17), 2216–2225.
20. Pohl, H. A. The Motion and Precipitation of Suspensoids in Divergent Electric Fields. *J. Appl. Phys.* **1951**, *22* (7), 869–871.
21. Pethig, R. R. *Dielectrophoresis: Theory, Methodology and Biological Applications*; John Wiley & Sons, 2017.
22. Henslee, E. A. Review: Dielectrophoresis in CELL Characterization. *Electrophoresis* **2020**, *41* (21–22), 1915–1930.
23. Arnold, W. M.; Zimmermann, U. Electro-Rotation—Development of a Technique for Dielectric Measurements on Individual Cells and Particles. *J. Electrost.* **1988**, *21* (2–3), 151–191.
24. Lei, U.; Yang, C. Y.; Wu, K. C. Viscous Torque on a Sphere Under Arbitrary Rotation. *Appl. Phys. Lett.* **2006**, *89* (18).
25. Adekanmbi, E. O.; Srivastava, S. K. Dielectric Characterization of Bioparticles via Electrokinetics: The Past, Present, and the Future. *Appl. Phys. Rev.* **2019**, *6* (4).
26. Broche, L. M.; Labeed, F. H.; Hughes, M. P. Extraction of Dielectric Properties of Multiple Populations From Dielectrophoretic Collection Spectrum Data. *Phys. Med. Biol.* **2005**, *50* (10), 2267–2274.
27. Huang, Y.; Wang, X. B.; Becker, F. F.; Gascoyne, P. R. C. Membrane Changes Associated With the Temperature-Sensitive P85(gag-mos)-Dependent Transformation of Rat Kidney Cells as Determined by Dielectrophoresis and Electrorotation. *Biochim. Biophys. Acta* **1996**, *1282* (1), 76–84.
28. Burt, J. P. H.; Pethig, R.; Gascoyne, P. R. C.; Becker, F. F. Dielectrophoretic Characterization of Friend Murine Erythroleukaemic Cells as a Measure of Induced-Differentiation. *Biochim. Biophys. Acta* **1990**, *1034* (1), 93–101.
29. Gagnon, Z.; Gordon, J.; Sengupta, S.; Chang, H. C. Bovine Red Blood Cell Starvation Age Discrimination Through a Glutaraldehyde-Amplified Dielectrophoretic Approach With Buffer Selection and Membrane Cross-Linking. *Electrophoresis* **2008**, *29* (11), 2272–2279.
30. Adams, T. N. G.; Jiang, A. Y. L.; Vyas, P. D.; Flanagan, L. A. Separation of Neural Stem Cells by Whole Cell Membrane Capacitance Using Dielectrophoresis. *Methods* **2018**, *133*, 91–103.
31. Afshar, S.; Salimi, E.; Fazelkhah, A.; Braasch, K.; Mishra, N.; Butler, M.; Thomson, D. J.; Bridges, G. E. Progression of Change in Membrane Capacitance and Cytoplasm Conductivity of Cells During Controlled Starvation Using Dual-Frequency DEP Cytometry. *Anal. Chim. Acta* **2019**, *1059*, 59–67.
32. Castellarnau, M.; Errachid, A.; Madrid, C.; Juarez, A.; Samitier, J. Dielectrophoresis as a Tool to Characterize and Differentiate Isogenic Mutants of *Escherichia coli*. *Biophys. J.* **2006**, *91* (10), 3937–3945.

33. Labeed, F. H.; Lu, J.; Mulhall, H. J.; Marchenko, S. A.; Hoettges, K. F.; Estrada, L. C.; Lee, A. P.; Hughes, M. P.; Flanagan, L. A. Biophysical Characteristics Reveal Neural Stem Cell Differentiation Potential. *PLoS ONE* **2011**, *6* (9), e25458.
34. Liang, W.; Zhao, Y.; Liu, L.; Wang, Y.; Li, W. J.; Lee, G. B. Determination of Cell Membrane Capacitance and Conductance via Optically Induced Electrokinetics. *Biophys. J.* **2017**, *113* (7), 1531–1539.
35. Oblak, J.; Krizaj, D.; Amon, S.; Macek-Lebar, A.; Miklavcic, D. Feasibility Study for Cell Electroporation Detection and Separation By Means of Dielectrophoresis. *Bioelectrochemistry* **2007**, *71* (2), 164–171.
36. Salmanzadeh, A.; Sano, M. B.; Gallo-Villanueva, R. C.; Roberts, P. C.; Schmelz, E. M.; Davalos, R.V. Investigating Dielectric Properties of Different Stages of Syngeneic Murine Ovarian Cancer Cells. *Biomicrofluidics* **2013**, 7 (1), 12.
37. Sano, M. B.; Henslee, E. A.; Schmelz, E. M.; Davalos, R.V. Contactless Dielectrophoretic Spectroscopy: Examination of the Dielectric Properties of Cells Found in Blood. *Electrophoresis* **2011**, *32* (22), 3164–3171.
38. Tada, S.; Eguchi, M.; Okano, K. Insulator-Based Dielectrophoresis Combined With the Isomotive AC Electric Field and Applied to Single Cell Analysis. *Electrophoresis* **2019**, *40* (10), 1494–1497.
39. Vykoukal, D. M.; Gascoyne, P. R.;Vykoukal, J. Dielectric Characterization of Complete Mononuclear and Polymorphonuclear Blood Cell Subpopulations for Label-Free Discrimination. *Integr. Biol. (Camb.)* **2009**, *1* (7), 477–484.
40. Zhou, T.; Ming, Y.; Perry, S. F.; Tatic-Lucic, S. Estimation of the Physical Properties of Neurons and Glial Cells Using Dielectrophoresis Crossover Frequency. *J. Biol. Phys.* **2016**, *42* (4), 571–586.
41. Unni, H. N.; Hartono, D.; Yung, L.Y. L.; Ng, M. M. L.; Lee, H. P.; Khoo, B. C.; Lim, K. M. Characterization and Separation of Cryptosporidium and Giardia Cells Using On-Chip Dielectrophoresis. *Biomicrofluidics* **2012**, *6* (1).
42. Wu, L.; Lanry Yung, L.Y.; Lim, K. M. Dielectrophoretic Capture Voltage Spectrum for Measurement of Dielectric Properties and Separation of Cancer Cells. *Biomicrofluidics* **2012**, *6* (1), 14113–1411310.
43. Gascoyne, P. R. C.; Noshari, J.; Becker, F. F.; Pethig, R. Use of Dielectrophoretic Collection Spectra for Characterizing Differences Between Normal and Cancerous Cells. *IEEE Trans. Ind. Appl.* **1994**, *30* (4), 829–834.
44. Hughes, M. P.; Morgan, H.; Rixon, F. J.; Burt, J. P. H.; Pethig, R. Manipulation of Herpes Simplex Virus Type 1 by Dielectrophoresis. *Biochim. Biophys. Acta* **1998**, *1425* (1), 119–126.
45. Labeed, F. H.; Coley, H. M.; Thomas, H.; Hughes, M. P. Assessment of Multidrug Resistance Reversal Using Dielectrophoresis and Flow Cytometry. *Biophys. J.* **2003**, *85* (3), 2028–2034.
46. Sanchis, A.; Brown, A. P.; Sancho, M.; Martinez, G.; Sebastian, J. L.; Munoz, S.; Miranda, J. M. Dielectric Characterization of Bacterial Cells Using Dielectrophoresis. *Bioelectromagnetics* **2007**, *28* (5), 393–401.
47. Gascoyne, P.; Pethig, R.; Satayavivad, J.; Becker, F. F.; Ruchirawat, M. Dielectrophoretic Detection of Changes in Erythrocyte Membranes Following Malarial Infection. *Biochim. Biophys. Acta* **1997**, *1323* (2), 240–252.
48. Huang, Y.; Wang, X. B.; Holzel, R.; Becker, F. F.; Gascoyne, P. R. Electrorotational Studies of the Cytoplasmic Dielectric Properties of Friend Murine Erythroleukaemia Cells. *Phys. Med. Biol.* **1995**, *40* (11), 1789–1806.
49. Keim, K.; Rashed, M. Z.; Kilchenmann, S. C.; Delattre, A.; Goncalves, A. F.; Ery, P.; Guiducci, C. On-Chip Technology for Single-Cell Arraying, Electrorotation-Based Analysis and Selective Release. *Electrophoresis* **2019**, *40* (14), 1830–1838.
50. Lannin, T.; Su, W. W.; Gruber, C.; Cardle, I.; Huang, C.; Thege, F.; Kirby, B. Automated Electrorotation Shows Electrokinetic Separation of Pancreatic Cancer Cells Is Robust

to Acquired Chemotherapy Resistance, Serum Starvation, and EMT. *Biomicrofluidics* **2016**, *10* (6), 064109.

51. Rohani, A.; Moore, J. H.; Su, Y. H.; Stagnaro, V.; Warren, C.; Swami, N. S. Single-Cell Electro-Phenotyping for Rapid Assessment of Clostridium Difficile Heterogeneity Under Vancomycin Treatment at Sub-MIC (Minimum Inhibitory Concentration) Levels. *Sensors Actuators B Chem.* **2018**, *276*, 472–480.
52. Liang, W.; Yang, X.; Wang, J.; Wang, Y.; Yang, W.; Liu, L. Determination of Dielectric Properties of Cells using AC Electrokinetic-based Microfluidic Platform: A Review of Recent Advances. *Micromachines (Basel)* **2020**, *11* (5).
53. Huang, Y.; Holzel, R.; Pethig, R.; Wang, X.-B. Differences in the AC Electrodynamics of Viable and Non-Viable Yeast Cells Determined Through Combined Dielectrophoresis and Electrorotation Studies. *Phys. Med. Biol.* **1992**, *37* (7), 1499–1517.
54. Park, S.; Capelin, D.; Piriatinskiy, G.; Lotan, T.; Yossifon, G. Dielectrophoretic Characterization and Isolation of Jellyfish Stinging Capsules. *Electrophoresis* **2017**, *38* (16), 1996–2003.
55. Sancho, M.; Martinez, G.; Martin, C. Accurate Dielectric Modelling of Shelled Particles and Cells. *J. Electrost.* **2003**, *57* (2), 143–156.
56. Markx, G. H.; Talary, M. S.; Pethig, R. Separation of Viable and Non-Viable Yeast Using Dielectrophoresis. *J. Biotechnol.* **1994**, *32* (1), 29–37.
57. Suehiro, J.; Hamada, R.; Noutomi, D.; Shutou, M.; Hara, M. Selective Detection of Viable Bacteria Using Dielectrophoretic Impedance Measurement Method. *J. Electrost.* **2003**, *57* (2), 157–168.
58. Zhang, J.; Song, Z.; Liu, Q.; Song, Y. Recent Advances in Dielectrophoresis-Based Cell Viability Assessment. *Electrophoresis* **2020**, *41* (10-11), 917–932.
59. Gascoyne, P. R.; Shim, S.; Noshari, J.; Becker, F. F.; Stemke-Hale, K. Correlations Between the Dielectric Properties and Exterior Morphology of Cells Revealed by Dielectrophoretic Field-Flow Fractionation. *Electrophoresis* **2013**, *34* (7), 1042–1050.
60. Wang, X. B.; Huang, Y.; Gascoyne, P. R. C.; Becker, F. F.; Holzel, R.; Pethig, R. Changes in Friend Murine Erythroleukemia Cell-Membranes During Induced-Differentiation Determined by Electrorotation. *Biochim. Biophys. Acta* **1994**, *1193* (2), 330–344.
61. Pethig, R.; Talary, M. S. Dielectrophoretic Detection of Membrane Morphology Changes in Jurkat T-cells Undergoing Etoposide-Induced Apoptosis. *IET Nanobiotechnol.* **2007**, *1* (1), 2–9.
62. Coley, H. M.; Labeed, F. H.; Thomas, H.; Hughes, M. P. Biophysical Characterization of MDR Breast Cancer Cell Lines Reveals the Cytoplasm Is Critical in Determining Drug Sensitivity. *Biochim. Biophys. Acta* **2007**, *1770* (4), 601–608.
63. Han, S.; Huang, C.; Han, A. Measurement of Dielectric Properties of Microalgae with Different Lipid Content Using Electrorotation and Negative Dielectrophoresis Cell Trap. In *2019 20th International Conference on Solid-State Sensors, Actuators and Microsystems & Eurosensors XXXIII (TRANSDUCERS & EUROSENSORS XXXIII)*; 2019.
64. Rohani, A.; Moore, J. H.; Kashatus, J. A.; Sesaki, H.; Kashatus, D. F.; Swami, N. S. Label-Free Quantification of Intracellular Mitochondrial Dynamics Using Dielectrophoresis. *Anal. Chem.* **2017**, *89* (11), 5757–5764.
65. Salimi, E.; Braasch, K.; Fazelkhah, A.; Afshar, S.; Saboktakin Rizi, B.; Mohammad, K.; Butler, M.; Bridges, G. E.; Thomson, D. J. Single Cell Dielectrophoresis Study of Apoptosis Progression Induced by Controlled Starvation. *Bioelectrochemistry* **2018**, *124*, 73–79.
66. Chen, J.; Xue, C.; Zhao, Y.; Chen, D.; Wu, M. H.; Wang, J. Microfluidic Impedance Flow Cytometry Enabling High-Throughput Single-Cell Electrical Property Characterization. *Int. J. Mol. Sci.* **2015**, *16* (5), 9804–9830.
67. Cheung, K. C.; Di Berardino, M.; Schade-Kampmann, G.; Hebeisen, M.; Pierzchalski, A.; Bocsi, J.; Mittag, A.; Tarnok, A. Microfluidic Impedance-Based Flow Cytometry. *Cytometry A* **2010**, 77 (7), 648–666.
68. Honrado, C.; Bisegna, P.; Swami, N.; et al. Single-Cell Microfluidic Impedance Cytometry: From Raw Signals to Cell Phenotypes Using Data Analytics. *Lab Chip* **2020**, *21*, 22–54.

69. Petchakup, C.; Li, K. H. H.; Hou, H.W. Advances in Single Cell Impedance Cytometry for Biomedical Applications. *Micromachines* **2017**, *8* (3).
70. Sun, T.; Morgan, H. Single-Cell Microfluidic Impedance Cytometry: A Review. *Microfluid. Nanofluid.* **2010**, *8* (4), 423–443.
71. De Ninno, A.; Reale, R.; Giovinazzo, A.; Bertani, F. R.; Businaro, L.; Bisegna, P.; Matteucci, C.; Caselli, F. High-Throughput Label-Free Characterization of Viable, Necrotic and Apoptotic Human Lymphoma Cells in a Coplanar-Electrode Microfluidic Impedance Chip. *Biosens. Bioelectron.* **2020**, *150*, 111887.
72. Gawad, S.; Cheung, K.; Seger, U.; Bertsch, A.; Renaud, P. Dielectric Spectroscopy in a Micromachined Flow Cytometer: Theoretical and Practical Considerations. *Lab Chip* **2004**, *4* (3), 241–251.
73. Daguerre, H.; Solsona, M.; Cottet, J.; Gauthier, M.; Renaud, P.; Bolopion, A. Positional Dependence of Particles and Cells in Microfluidic Electrical Impedance Flow Cytometry: Origin, Challenges and Opportunities. *Lab Chip* **2020**, *20* (20), 3665–3689.
74. Errico, V.; Ninno, A. D.; Bertani, F. R.; Businaro, L.; Bisegna, P.; Caselli, F. Mitigating Positional Dependence in Coplanar Electrode Coulter-Type Microfluidic Devices. *Sensors Actuators B Chem.* **2017**, *247*, 580–586.
75. Spencer, D.; Morgan, H. High-Speed Single-Cell Dielectric Spectroscopy. *ACS Sens.* **2020**, *5* (2), 423–430.
76. Honrado, C.; Ciuffreda, L.; Spencer, D.; Ranford-Cartwright, L.; Morgan, H. Dielectric Characterization of Plasmodium Falciparum-Infected Red Blood Cells Using Microfluidic Impedance Cytometry. *J. R. Soc. Interface* **2018**, *15* (147).
77. Bürgel, S. C.; Escobedo, C.; Haandbæk, N.; Hierlemann, A. On-Chip Electroporation and Impedance Spectroscopy of Single-Cells. *Sensors Actuators B Chem.* **2015**, *210*, 82–90.
78. Zhang, Y.; Liang, H.; Tan, H.; Chen, D.; Wang, Y.; Xu, Y.; Wang, J.; Chen, J. Development of Microfluidic Platform to High-Throughput Quantify Single-Cell Intrinsic Bioelectrical Markers of Tumor Cell Lines, Subtypes and Patient Tumor Cells. *Sensors Actuators B Chem.* **2020**, *317*.
79. Caselli, F.; De Ninno, A.; Reale, R.; Businaro, L.; Bisegna, P. A Novel Wiring Scheme for Standard Chips Enabling High-Accuracy Impedance Cytometry. *Sensors Actuators B Chem.* **2018**, *256*, 580–589.
80. Spencer, D.; Caselli, F.; Bisegna, P.; Morgan, H. High Accuracy Particle Analysis Using Sheathless Microfluidic Impedance Cytometry. *Lab Chip* **2016**, *16* (13), 2467–2473.
81. Spencer, D. C.; Paton, T. F.; Mulroney, K. T.; Inglis, T. J. J.; Sutton, J. M.; Morgan, H. A Fast Impedance-Based Antimicrobial Susceptibility Test. *Nat. Commun.* **2020**, *11* (1), 5328.
82. De Ninno, A.; Errico, V.; Bertani, F. R.; Businaro, L.; Bisegna, P.; Caselli, F. Coplanar Electrode Microfluidic Chip Enabling Accurate Sheathless Impedance Cytometry. *Lab Chip* **2017**, *17* (6), 1158–1166.
83. Yang, D. H.; Ai, Y. Microfluidic Impedance Cytometry Device With N-Shaped Electrodes for Lateral Position Measurement of Single Cells/Particles. *Lab Chip* **2019**, *19* (21), 3609–3617.
84. Demierre, N.; Braschler, T.; Linderholm, P.; Seger, U.; van Lintel, H.; Renaud, P. Characterization and Optimization of Liquid Electrodes for Lateral Dielectrophoresis. *Lab Chip* **2007**, 7 (3), 355–365.
85. Shaker, M.; Colella, L.; Caselli, F.; Bisegna, P.; Renaud, P. An Impedance-Based Flow Microcytometer for Single Cell Morphology Discrimination. *Lab Chip* **2014**, *14* (14), 2548–2555.
86. Wang, N.; Liu, R.; Asmare, N.; Chu, C. H.; Sarioglu, A. F. Processing Code-Multiplexed Coulter Signals via Deep Convolutional Neural Networks. *Lab Chip* **2019**, *19* (19), 3292–3304.

87. Rollo, E.; Tenaglia, E.; Genolet, R.; Bianchi, E.; Harari, A.; Coukos, G.; Guiducci, C. Label-Free Identification of Activated T Lymphocytes Through Tridimensional Microsensors On Chip. *Biosens. Bioelectron.* **2017**, *94*, 193–199.
88. Clausen, C.; Skands, G.; Bertelsen, C.; Svendsen, W. Coplanar Electrode Layout Optimized for Increased Sensitivity for Electrical Impedance Spectroscopy. *Micromachines* **2014**, *6* (1), 110–120.
89. Cottet, J.; Kehren, A.; van Lintel, H.; Buret, F.; Frénéa-Robin, M.; Renaud, P. How to Improve the Sensitivity of Coplanar Electrodes and Micro Channel Design in Electrical Impedance Flow Cytometry: A Study. *Microfluid. Nanofluid.* **2019**, *23* (1).
90. Ghassemi, P.; Harris, K. S.; Ren, X.; Foster, B. M.; Langefeld, C. D.; Kerr, B. A.; Agah, M. Comparative Study of Prostate Cancer Biophysical and Migratory Characteristics via Iterative Mechanoelectrical Properties (iMEP) and Standard Migration Assays. *Sensors Actuators B Chem.* **2020**, *321*.
91. Yang, D.; Zhou, Y.; Zhou, Y.; Han, J.; Ai, Y. Biophysical Phenotyping of Single Cells Using a Differential Multiconstriction Microfluidic Device With Self-Aligned 3D Electrodes. *Biosens. Bioelectron.* **2019**, *133*, 16–23.
92. Weiz, S. M.; Medina-Sánchez, M.; Schmidt, O. G. Microsystems for Single-Cell Analysis. *Adv. Biosyst.* **2018**, *2* (2).
93. Reale, R.; De Ninno, A.; Businaro, L.; Bisegna, P.; Caselli, F. Electrical Measurement of Cross-Sectional Position of Particles Flowing Through a Microchannel. *Microfluid. Nanofluid.* **2018**, *22* (4).
94. McGrath, J. S.; Honrado, C.; Moore, J. H.; Adair, S. J.; Varhue, W. B.; Salahi, A.; Farmehini, V.; Goudreau, B. J.; Nagdas, S.; Blais, E. M.; Bauer, T. W.; Swami, N. S. Electrophysiology-Based Stratification of Pancreatic Tumorigenicity by Label-Free Single-Cell Impedance Cytometry. *Anal. Chim. Acta* **2020**, *1101*, 90–98.
95. Fung, C. W.; Chan, S. N.; Wu, A. R. Microfluidic Single-Cell Analysis-Toward Integration and Total On-Chip Analysis. *Biomicrofluidics* **2020**, *14* (2), 021502.
96. Galan, E. A.; Zhao, H.; Wang, X.; Dai, Q.; Huck, W. T. S.; Ma, S. Intelligent Microfluidics: The Convergence of Machine Learning and Microfluidics in Materials Science and Biomedicine. *Matter* **2020**, *3* (6), 1893–1922.
97. Isozaki, A.; Harmon, J.; Zhou, Y.; Li, S.; Nakagawa, Y.; Hayashi, M.; Mikami, H.; Lei, C.; Goda, K. AI on a chip. *Lab Chip* **2020**, *20* (17), 3074–3090.
98. Riordon, J.; Sovilj, D.; Sanner, S.; Sinton, D.; Young, E. W. K. Deep Learning with Microfluidics for Biotechnology. *Trends Biotechnol.* **2019**, *37* (3), 310–324.
99. Dudaie, M.; Nissim, N.; Barnea, I.; Gerling, T.; Duschl, C.; Kirschbaum, M.; Shaked, N. T. Label-Free Discrimination and Selection of Cancer Cells From Blood During Flow Using Holography-Induced Dielectrophoresis. *J. Biophotonics* **2020**, *13* (11), e202000151.
100. Alizadeh-Haghighi, E.; Jafarmadar, S.; Khalilarya, S. Application of Genetic Algorithm in Extracting Cell Dielectric Characteristics With Electrorotation. *J. Electr. Bioimpedance* **2019**, *8* (1), 34–39.
101. Zhao, Y.; Wang, K.; Chen, D.; Fan, B.; Xu, Y.; Ye, Y.; Wang, J.; Chen, J.; Huang, C. Development of Microfluidic Impedance Cytometry Enabling the Quantification of Specific Membrane Capacitance and Cytoplasm Conductivity From 100,000 Single Cells. *Biosens. Bioelectron.* **2018**, *111*, 138–143.
102. Honrado, C.; McGrath, J. S.; Reale, R.; Bisegna, P.; Swami, N. S.; Caselli, F. A Neural Network Approach for Real-Time Particle/Cell Characterization in Microfluidic Impedance Cytometry. *Anal. Bioanal. Chem.* **2020**, *412* (16), 3835–3845.

CHAPTER 5

Biosensors for single-cell mechanical characterization

Minhui Liang, Jianwei Zhong, and Ye Ai*
Pillar of Engineering Product Development, Singapore University of Technology and Design, Singapore, Singapore

1 Introduction

Cell mechanical phenotypes have attracted great attention as a new type of label-free biomarker to indicate cell states or diseases. A label-free biomarker has plentiful merits compared with a molecular biomarker, such as cost reduction without label reagent, time-saving on labor-intensive labeling and washing steps, and improved repeatability with direct sample measurement. The mechanical phenotyping of individual biological cells is powerful and unique. Substantial fundamental studies have linked cell mechanical phenotypes with cellular components, such as cytoskeleton[1–4] and nucleus.[5–7] Various examples have shown that the stiffness changes can indicate cell states such as the activation of leukocytes,[8, 9] malignancy of cancer cells,[10–13] or disease like malaria.[8, 14, 15] In this chapter, we discuss the origins of cell mechanics and follow several examples of how cell mechanical phenotyping is relevant to clinical applications. Then, we will go through the conventional techniques and highlight the current microfluidics-based approaches which will bring cell mechanical phenotyping to a promising future.

2 Biology background

Cell mechanical phenotypes are complex as a synergy of various subcellular components contributing differently (Fig. 1). The mechanical properties of a single cell are dominated by the function of the nucleus and cytoplasm.[16, 17] The nucleus, particularly, raises interest from researchers because it is the largest and stiffest organelle.[6, 7] For example, Xia et al.[18] and Rowat et al.[19] discovered that the nucleus deformability of circulating tumor cells and neutrophils plays an essential role in the ability of the cell to translocate in narrow constrictions. Researchers have further studied the nucleus stiffness into protein level.

*Corresponding author.

https://doi.org/10.1016/B978-0-323-89841-6.00006-2

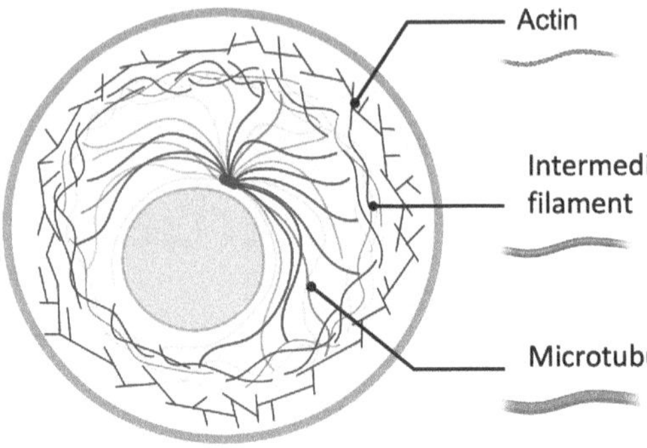

Fig. 1 Schematic of cellular components that contribute to cell mechanical responses. *(Adapted from "Cytoskeleton Components," by BioRender.com (2021). Retrieved from https://app.biorender.com/biorender-templates).*

As an elementary protein composing the inner nuclear membrane, lamin A contributes mainly to nucleus stability and is frequently modified for nuclear deformability studies.[19–21] For example, the nuclear shape was regarded as an essential indicator for the deformability of neutrophils. However, Rowat et al.[19] drew a different conclusion that the nuclear shape could be less critical than the level of lamin A in cell mechanical phenotypes. In this report, the low lamin A level is deduced to be the primary cause of cell deformability changes, and it, in turn, results in an irregular nuclear shape.

Additionally, the other major contributor to cell mechanics is the cytoskeleton within the cytoplasm. It maintains a cell structure, organelle position, providing resistance to external forces. Among the components in the cytoskeleton, actin, microtubules, and intermedia filaments are the major contributors. Each of the components has a specific role based on rheology differences.[17] For example, actin organizes the position of the organelles and may be fluidized at high strain to provide rapid deformability; microtubules work for transportation and cellular stabilization; the intermedia filaments support cellular integrity under high strain.[22] To analyze a single component, drugs are frequently used to stabilize or disrupt the polymerization process for changing cell mechanical properties. For instance, cytochalasin D is a common drug for cellular actin disruption, which results in a cell deformability change.[4] Titushkin[23] applied cytochalasin D to disrupt actin polymerization for examining the cell membrane-cytoskeleton associated mechanical phenotype. This work has demonstrated that the cellular membrane has a relatively weak impact on the mechanical response because of its high extensibility.

3 Importance of sing-cell mechanical phenotyping

Revealing the underlying mechanism of cell mechanical responses has paved the way for vast applications using cell mechanical phenotyping as a label-free indicator. For instance, the store duration of RBC could induce RBC biochemical and functional changes, which might increase the risk of RBC transfusion.[24–26] Among the changes of RBC properties, the decrease of deformability can be employed as a simple and reliable label-free indicator. In early 1969, Weed et al.[27] reported the RBC deformability decreases with ATP depleting using a micropipette. Another work applied ektacytometry for RBC deformability measurement showing the deformability decreases with longer storage duration.[24] More recently, a microfluidic method was performed to investigate the RBC deformability in a human-capillary-like channel.[28] In this work, the RBC deformation shape (circularity) and the recovery from the deformation (recovery time) were investigated to indicate RBC deformability. It has been found that the aspect ratio of RBC deformation (deformability = the major axis length/the minor axis length) did not reveal a stiffness variation with different storage durations. Deformability fails to show the RBC stiffness changes may be because the RBC has various spectrum of motions[29–31] when transiting in microchannels under different flow conditions. RBC can appear in shapes of tumbling, parachute, and tank-treading,[29–31] interfering shape measurements. More recently, Saadat et al.[32] chose specific flow conditions that first guarantee a stable parachute shape during the transition of RBC. Then, indicating the RBC stiffening (become less deformable) by Taylor deformation parameter ($\frac{a-b}{a+b}$, where a is the major axis length and b is the minor axis length of deformed RBC), which is stated to be superior to circularity or the recovery time for indicating RBC stiffness sensitively. Substantial evidence has suggested the RBC deformability is related to RBC quality. In turn, RBC deformability has been demonstrated in many reports as a label-free indicator for RBC quality during storage.

Another well-known example is cancer cells have much lower elastic stiffness than normal cells, and it is believed that malignant cells are softer than nonmalignant cells.[33] Substantial results indicated the cellular constituents involved in cancer development,[34] and proved the high deformability of cancer cells is an important indicator useful for label-free cancer screening.[33] For instance, Rother[11] revealed the mechanical phenotyping differences between stiffer benign MCF-10A and softer malignant

MDA-MB-231. Xu et al.[10] illustrated that ovarian cancer has lower stiffness than nonmalignant ovarian epithelial cells. Lekka et al.[12] showed that most cancerous cell lines (such as bladder, melanoma, prostate, breast, and colon cancer cells) are more deformable comparing with their nontumorigenic or nonmalignant or less differentiated cancers cells. Furthermore, Darling et al.[35] presented the variation of deformability in human chondrosarcoma cell lines of different differentiation and malignancy level. The results suggest that the cell mechanical characteristics may reflect a metastatic cellular process.[35] More than distinguishing malignant or non-malignant cancer cells, Swaminathan et al.[13] raised the resolution of cellular mechanical phenotypes to grade the potential metastatic level. In this work, an inverse power-law relationship of metastatic potential and cell stiffness was built, providing further insights into cancer progression. To the protein level, researchers have given insight into the impact of cytoskeleton organization on the mechanical properties, especially the actin microfilament and microtubules. Compared with normal cells in which the actin microfilament contributes more to cell elasticity, the microtubules appear to be more important for cancer cells with respect to the mechanical response.[1] Furthermore, Pachenari[2] stated the ratio of microtubules to actin microfilament and the concentration of actin filament around the cell membrane indicates a cancer transition from nonaggressive to malignant. The reorganization in the cytoskeleton facilitates the deformability of cancer cells that being squeezed through microvasculature. Many anticancer drugs act by modifying cytoskeleton.[36] Thus, these findings propel the development of pharmacologic approaches to regulate cancer metastasis by suppressing the migration and invasion through controlling the mechanical properties.[37]

To conclude, the synergy of cytoskeleton and nucleus in single cells results in mechanical responses can indicate cell states and diseases. More specifically, the nucleus contributes as the stiffest and largest organelle, and the cytoskeleton consists of diverse components that each attains a different role for maintaining interior and exterior cell structure. After knowing the underlying mechanism of cell mechanical response, we make two examples of using this mechanical phenotype as a label-free marker for clinical and research use: The first example is RBC mechanical property changes during storage. The increase of stiffness is used as indicators for RBC transfusion quality. The other example is the decrease of stiffness observed in cancer cells and further stiffness changes in the metastatic cancer process. Moreover, the changes of cancer cell mechanical property at the cellular organelle level are related to the reorganization of microfilaments and microtubules, which may have the potential to propel anti-cancer drug studies. Taken together, cell mechanical phenotyping is highly valuable to serve as a label-free biomarker in research or clinical diagnosis. In the next section, we will discuss the approaches for characterizing cell mechanical phenotypes.

4 Conventional methods

Over the decades, cell mechanical phenotyping, also termed mechanophenotyping, has been studied with a variety of tools. The well-developed conventional tools offer high precision results but are limited by the throughput. Because of the cell heterogeneity, the cell properties vary in a certain range, such as the cell size may appear in a normal or log-normal distribution. The sample volume should be comparable with the flow cytometry (hundreds to thousands of cells analyzed per run) to be statistically meaningful. For example, micropipette aspiration (MA) is a widely applied tool dating back to the 1990s.[38–40] It is a method that employs suction pressure in a small glass tube to investigate the response of a single cell in the glass tube by microscope (Fig. 2A). This process involves highly skilled and cumbersome manual operation is time-consuming and labor-intensive. Nevertheless, MA has provided precise mechanical information for cell lines such as leukocytes,[41,42] endothelial cells,[43] and erythrocytes.[44,45] Another remarkable contribution of MA is that it inspired microfluidics-based constriction deformability cytometry, which will be discussed later. In contrast to MA which is specialized for measuring nonadherent cells (cells suspended in a solution), an atomic force microscope (AFM) measures adherent cells (i.e., cells attached to a surface). AFM is a precise tool with atomic resolution widely applied in a broad spectrum of field such as materials, electronics, and biology.[10,46] This system consists of a cantilever to apply force onto the cell, causing an indentation measured through laser deflection and collected by a photodetector (Fig. 2B). Same as MA, AFM requires skilled technicians and is time-consuming. However, AFM serves as an indispensable tool allowing a point-by-point measurement, which can reveal the mechanical properties of a specific part of the cell. For instance, AFM provides evidence that the nucleus is the stiffest organelle and gives insight into the contribution of the individual nuclear envelope to nuclear mechanics.[47] On the other hand, it comes with a problem that Young's Modulus of the cell measured by AFM varies widely depending on multiple factors such as the cantilever tip, probing region, measurement parameter, or substrate stiffness.[48,49] Because of the highly

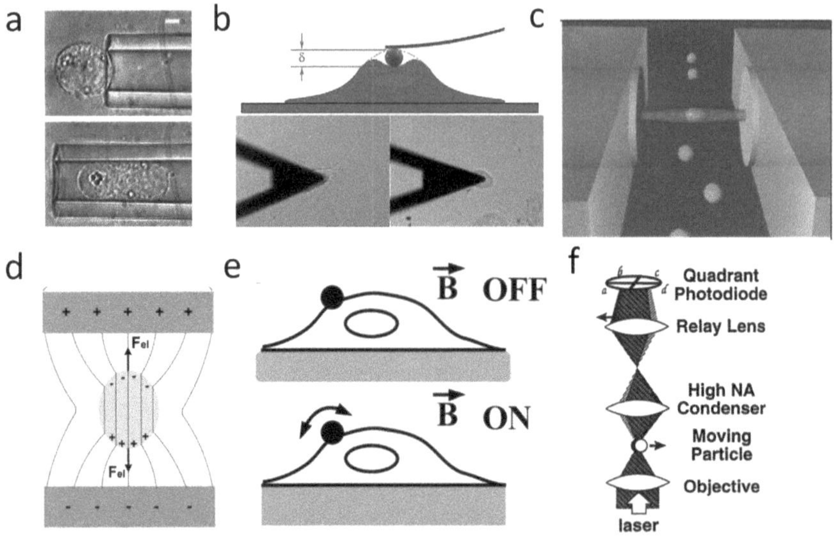

Fig. 2 Conventional methods for cell mechanical phenotyping assessment. (A) Micropipette aspiration. (B) Atomic force microscopy. (C) Optical stretcher. (D) Dielectrophoresis method. (E) Magnetic twisting cytometer. (F) Particle tracking microrheology. *(Part A: Jones, W. R.; Ting-Beall, H. P.; Lee, G. M.; Kelley, S. S.; Hochmuth, R. M.; Guilak, F. Alterations in the Young's modulus and volumetric properties of chondrocytes isolated from normal and osteoarthritic human cartilage. J. Biomech.* ***1999,*** *32, 119–127, Copyright 1999 Elsevier Science Ltd. Part B: Xu, W.; Mezencev, R.; Kim, B.; Wang, L.; McDonald, J.; Sulchek, T. Cell stiffness is a biomarker of the metastatic potential of ovarian cancer cells. PLoS One* ***2012,*** *7, e46609, Copyright Xu et al. under the terms of the Creative Commons Attribution License. Part C: Guck, J.; Schinkinger, S.; Lincoln, B.; Wottawah, F.; Ebert, S.; Romeyke, M.; Lenz, D.; Erickson, H. M.; Ananthakrishnan, R.; Mitchell, D.; Käs, J.; Ulvick, S.; Bilby, C. Optical deformability as an inherent cell marker for testing malignant transformation and metastatic competence. Biophys. J.* ***2005,*** *88, 3689–3698, Copyright 2005 The Biophysical Society. Part D: Teng, Y.; Pang, M.; Huang, J.; Xiong, C. Mechanical characterization of cancer cells during TGF-β1-induced epithelial-mesenchymal transition using an electrode formation-based microchip. Sens. Act. Chem.* ***2017,*** *240, 158–167, Copyright 2016 Elsevier B.V. Part E: Massiera, G.; Citters, K. M. V.; Biancaniello, P. L.; Crocker, J. C. Mechanics of single cells: Rheology, time dependence, and fluctuations. Biophys. J.* ***2007****, 93, 3703–3713, Copyright 2007, The Biophysical Society. Part F: Yamada, S.; Wirtz, D.; Kuo, S. C. Mechanics of living cells measured by laser tracking microrheology. Biophys. J.* ***2000,*** *78, 1736–1747, Copyright 2000 The Biophysical Society.)*

manual operation dependency, AFM and MA have a very low throughput of < 1 cell per min.[36,48]

More recently, approaches to characterize cell stiffness in suspension by applying force generated from optics and dielectrophoresis have been developed. An optical tweezer (OT) is a well-studied method that uses a high-power laser beam to generate an optical pressure gradient on the cell surface (Fig. 2C).[50, 51] The light transfers momentum on the cell surface, which results in opposite forces to stretch the cell. In highly inhomogeneous electric fields, charges are trapped on the cell surface, leading to a cell polarization that tends to elongate it (Fig. 2D).[52, 53] However, these methods require highly intensive power, which may cause heating and damage to cells, and the throughput is low (~1 cell per min). Because of a small amount of force exerted on cells, it requires energy accumulation for cell stretching to a maximum deformation. Another type of technique, magnetic twisting cytometer (MTC) and particle tracking microrheology (PTM), is able to give insight into cell frequency-dependent rheology information. In MTC, an external magnetic bead is bounded to the surface of a single cell, then MTC manipulates the bead with an oscillatory magnetic field and measures the corresponding response of cells (Fig. 2E).[54] PTM tracks the Brownian motion with fluorescent beads that are injected into cells. The magnitude and time scale of the particle's trajectory is used to provide cellular mechanical information (Fig. 2F).[3] The drawbacks of MTC and PTM are label-required, low throughput (~30 cell per min for MTC, <1 cell per min for PTM) and the magnitude of the applied force is small (piconewton).[16,48]

It is remarkable that the conventional cell mechanical phenotyping methods have been compared systematically by Wu.[48] The measured Yong's modulus of MCF-7 is widely distributed within and between methods. For example, the results obtained by AFM vary more than tenfold; the measured results of MTC are >100-fold higher than PTM, while at the same range compared with AFM results.[48] A primary reason is the diversity of cellular response to different force profiles. For example, AFM and MTC measure with a tiny contact area of a compressive force, while MA and OS use immense forces. MTC can exert oscillated forces, while MA and OS apply continuous forces. AFM can employ a more considerable force magnitude (micronewton), while MTC and PTM only exert a small magnitude of force (piconewton).[16,48] These diverse approaches allow researchers to choose the investigation method according to their needs. However, the flip side of the coin is that the lack of standardization imposes a considerable challenge for interpreting and comparing results with different methods mentioned by Guck[55] and Carlo[36]. Besides, a critical limitation for conventional cell mechanical phenotyping approaches is the low throughput. Due to the heterogeneity within a large cell population or mixed samples, a large number of individual cells should be measured to reveal the rare cell types and reach statistical confidence. To address this challenge, substantial microfluidic approaches have emerged, which we will discuss in detail in the following section.

5 Microfluidics-based methods

The bright prospects of single-cell mechanics have elicited a flurry of microfluidics-based measuring systems. The microfluidic system provides a throughput of up to thousands of cells per second. Furthermore, suspended cells in a flow translocating in a capillary-like channel offer a similar scene in vivo. The microfluidic method integrated with electrical, optical, imaging measurements is highly automatic. The minimization of manual intervention could decrease errors, which in turn increase the reliability and repeatability. Recently, there are three classes of microfluidics-based cell mechanical phenotyping methods. One primary category is the noncontactless method that applies a channel with a constriction smaller than cell size. The other two classes are contactless methods that employ hydrodynamic forces to stretch cells. This section summarizes and discusses the state of art methods and their applications.

5.1 Constriction deformability cytometry

The ability of cells to traverse through a narrow constriction is critical for cell processes from immune response to metastasis. Such as, activated leukocytes are required to migrate to sites of infection[56]; malignant cancer cells are softer to propel invasiveness [33]; the malaria-infected RBCs become stiffer and in turn, contribute to capillary blockage.[14] As the transition of cells passing through a constriction is vital for in vivo functioning and useful in indicating the mechanical properties, researchers have developed many methods to investigate single-cell transit capability. Conceived from the idea of MA, microconstriction is integrated with microfluidic techniques to achieve higher throughput. A precise approach measures the transit time of the cell through the constriction by a suspended microchannel resonator (SMR, Fig. 3A). It provides multiple information, including size, entry velocity (velocity before entering the constriction), transit velocity (velocity in the constriction), passage time, and deformability for each cell event.[57–59] Furthermore, Byun et al.[58] suggested the entry velocity mainly indicates cell stiffness, whereas transit velocity indicates the synergy of cell stiffness and friction between the cell surface and channel wall. Because the process of cell entering the constriction involves less cell-wall contact than the process of cell transiting in the constriction, therefore, in this work, the

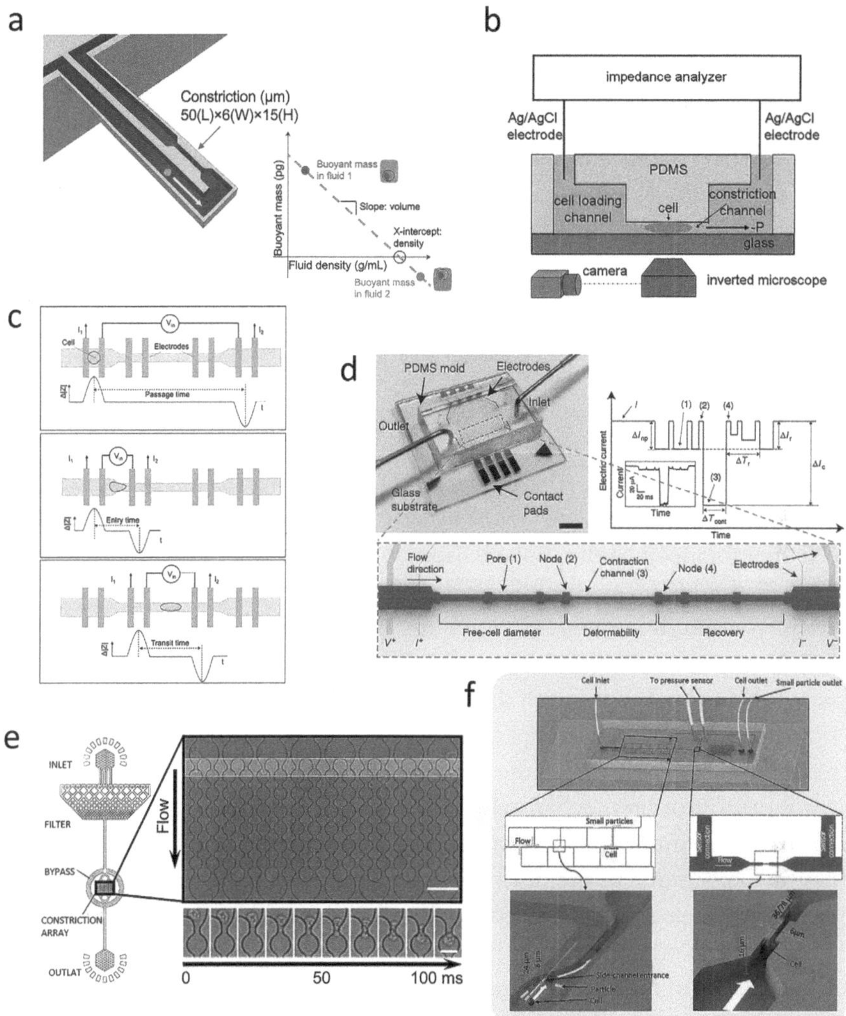

Fig. 3 Representative approaches in constriction deformability cytometry (cDC) for cell mechanical phenotyping assessment. (A) Suspended microchannel resonator. (B) Impedance-based cDC. (C) Coplanar differential impedance sensing. (D) Mechanical node-pore sensing. (E) Imaging-based quantitative cDC. (F) Pressure feedback cDC. *(Part A: Byun, S.; Hecht, V. C.; Manalis, S. R. Characterizing cellular biophysical responses to stress by relating density, deformability, and size. Biophys. J.* ***2015,*** *109, 1565–1573, Copyright 2015 Biophysical Society. Part B: Chen, J.; Zheng, Y.; Tan, Q.; Shojaei-Baghini, E.; Zhang, Y. L.; Li, J.; Prasad, P.; You, L.; Wu, X. Y.; Sun, Y. Classification of cell types using a microfluidic device for mechanical and electrical measurement on single cells. Lab Chip* ***2011,*** *11, 3174–3181, Copyright 2011 The Royal Society of Chemistry. Part C: Zhou, Y.; Yang, D.; Zhou, Y.; Khoo, B. L.; Han, J.; Ai, Y. Characterizing deformability and electrical impedance of cancer cells in a*
(Continued)

comparison of entry velocity with transit velocity could suggest reduced friction appears in invasive cancer cells.

Date back to the 1990s, Fisher[60] has described a cell transit analyzer that measures the transit time when cells pass through the pore by detecting the "resistive pulse." This prototype is a branch of impedance cytometry[61] that electrodes are placed at each side of the constriction applying a voltage to detect impedance changes. When a cell passes through the sensing region, it replaces the conductive medium causing an impedance change (Fig. 3B–D). Chen et al.[62] applied this idea and reported the pulse width and amplitude of impedance signal when a cell passes through the constriction provide both mechanical and electrical information. Combining mechanical and electrical phenotyping enhances the cell classification success rate between cell lines compared with using a single indicator alone (Fig. 3B). Furthermore, the impedance signal provides rich information at different frequency regions. More specifically, the impedance could reflect cell membrane-medium polarization at a low frequency while reflecting the cell cytoplasm-medium polarization when the cell membrane is short-circuited at high frequency.[61] Applying multifrequency measurement, both cell membrane capacitance and the cytoplasm resistance can be obtained when the cell passes through the detection area.[63] Additionally, the constriction provides another two parameters to give insight into cell mechanical response from the transit time and cell elongation in the constriction. The combination of biomechanical and bioelectrical measurement is proved to be useful for improving the success cell classification rate.[62] Adamo et al.[64] has further simplified the device by using the coplanar electrodes. More recently, Zhou et al.[65] applied two pairs of differential coplanar electrodes to obtain entry and transit time (Fig. 3C). Meanwhile, the differential configuration is able to cancel out signal drifting like the fluctuation of temperature,

---

Fig. 3, cont'd *microfluidic device. Anal. Chem.* ***2018,*** *90, 912–919, Copyright 2018 American Chemical Society. Part D: Kim, J.; Han, S.; Lei, A.; Miyano, M.; Bloom, J.; Srivastava, V.; Stampfer, M. R.; Gartner, Z. J.; LaBarge, M. A.; Sohn, L. L. Characterizing cellular mechanical phenotypes with mechano-node-pore sensing. Microsyst. Nanoeng.* ***2018,*** *4, 17091, Copyright 2018 The Authors under the terms of the Creative Commons CC BY license. Part E: Nyberg, K. D.; Hu, K. H.; Kleinman, S. H.; Khismatullin, D. B.; Butte, M. J.; Rowat, A. C. Quantitative deformability cytometry: Rapid, calibrated measurements of cell mechanical properties. Biophys. J.* ***2017,*** *113, 1574–1584, Copyright 2016 The Royal Society of Chemistry. Part F: Chen, Z.; Zhu, Y.; Xu, D.; Alam, Md. M.; Shui, L.; Chen, H. Cell elasticity measurement using a microfluidic device with real-time pressure feedback. Lab Chip* ***2020,*** *20, 2343–2353, Copyright 2020 The Royal Society of Chemistry.)*

pH, or conductivity. In this work, passage time (the total time cell passes through two pairs of electrodes), entry time, and transit time show the same trend, decreasing with the stiffness. Moreover, the comparison of impedance amplitude of deformed (cell in the constriction) and undeformed (before entering the constriction) cells demonstrates a connection between impedance amplitude and cellular mechanical properties. For example, the impedance amplitude of fixed MCF-7 is the least as compared with undeformed normal and chemically treated MCF-7, while it seems the largest when deformed. A hypothesis is that the impedance amplitude is determined by the contact of the cell membrane to the channel wall and cell membrane capacitance when the cell seals the channel. Thus fixed cell seals the channel the most because of the largest stiffness, which could reflect the mechanical properties by impedance amplitude.

Based on the electrical-based constriction system, by introducing Node and Pore (Fig. 3D) to the channel configuration, Kim et al.[66] characterized cellular recovery time, an additional measuring dimension, to the system. Recovery time provides unique information for a cell population. Kim showed that recovery is a distinct character related to cellular actin filament concentration and chronological age. In Yang's work,[67] relaxation of cells is inspected by the transit time of cells passing through multiconstriction. The faster the cell is recovered to a sphere shape, the more time it takes to traverse through the next constriction, vice versa. Moreover, the Node and Pore design can reveal the storage (elasticity) and loss (viscosity) moduli of cells. Thus it may further determine the individual contributions of cytoskeleton components such as actin and microtubules stated by Kim et al.[68] The electrical-based cDC provides both bioelectrical and biomechanical properties cells at a high throughput of up to hundreds of cells per minute.[65] Moreover, the Node and Pore channel offers cell recovery measurement. The impedance signal has merits for information interpreting compared with the imaging-based method we will discuss next. For example, it requires less computational power, less memory space, offers real-time capability.

Imaging is the other widely used approach to investigate dynamic cell motion through constriction. Comparing with the electrical-based method, the imaging-based approach requires more computational power. But the imaging-based method allows higher throughput by scanning an array of constrictions at the same time. Nyberg et al.[69] and Lange et al.[21] reported methods for quantitative measurements reaching a throughput of up to 1000 cells per min. They applied an array of short constriction and record the translocation of cells by a high-speed camera (Fig. 3E), a model to interpret Young's modulus and fluidity of cells was built. Furthermore, the contribution of individual cellular components was investigated. Besides, vast applications and basic studies developed with an array-like channel to investigate cell mechanical properties.[15, 18, 19, 70–72]

In cDC, pressure feedback is another approach to gain insight into cell mechanical response. Chen et al.[73] reported a microfluidics system combining a size-based separation unit and a precise differential pressure sensor to collect pressure feedback when cells squeeze through the constriction (Fig. 3F). Early in 2009, the pressure differentiation was read by the height difference of PBS in burettes.[74] However, this device is bulky and complex to fabricate.

Taken together, the diversity of cDC is remarkable. Moreover, it is a simple, high-throughput, low-cost method for single-cell mechanical phenotyping. However, clogging is a major limitation that hinders cDC from being a wide-spread approach. The other restriction emerges when investigating the cell response in the constriction. Factors that determine the transit time like cell deformability, friction, and size are not completely decoupled. First, cell transit time in the constriction strongly depends on cell size.[21, 58, 75] Second, cell stiffness affects passage time.[21, 58, 75] The less deformable cells cause a larger normal force against the channel wall, which increases the friction. Third, the roughness of the cell surface within or between cell lines is no constant. Even the roughness and hydrophility of the channel wall is variable.[75] In other words, the cell dynamic in constriction is a complex process related to entangled parameters. Thus the entry velocity, recovery time, and cell extension length are investigated together with the cell transit time in the constriction to evaluate the cell mechanical properties.[58, 75, 76] Because of the physical contact between cell and channel wall, intricated factors emerged, which imposed a considerable challenge for deformability interpreting. Thus the contactless method is elicited to address these challenges.

5.2 Shear flow deformability cytometry

In the recent years, the contactless hydrodynamic stretching method has emerged for single-cell mechanical phenotyping with an unprecedented throughput that reaches up to thousands of cells per second. This method can be classified into two groups based on the flow direction. One major group, shear flow deformability cytometry (sDC) applies constriction that is slightly larger than cell size. Cells migrate with the straight flow streamline without touching the channel wall (Fig. 4A). The shear stress acting

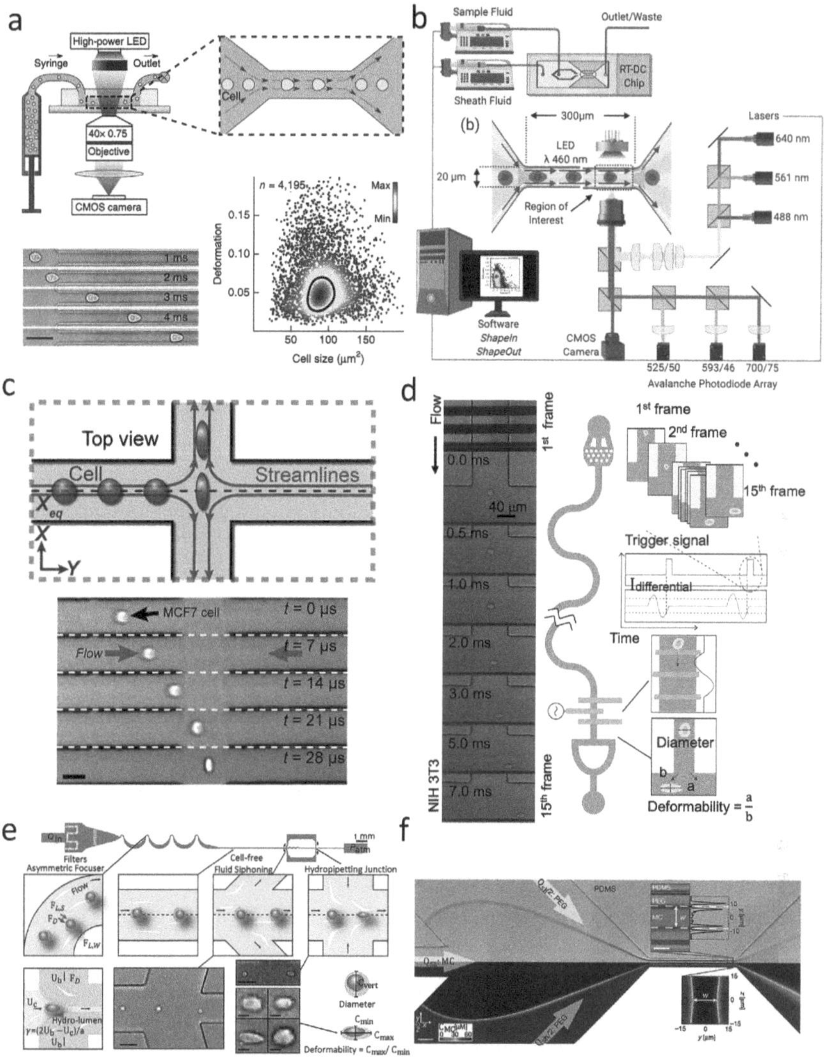

Fig. 4 Representative approaches in hydrodynamic deformability cytometry for cell mechanical phenotyping assessment. (A) Real-time deformability cytometry. (B) Real-time fluorescence deformability cytometry. (C) Deformability cytometry. (D) Electronically triggered deformability cytometry with on-demand imaging in viscoelastic fluids. (E) Pinched-flow hydrodynamic stretching. (F) Virtual fluidic channel deformability cytometry. *(Part A: Otto, O.; Rosendahl, P.; Mietke, A.; Golfier, S.; Herold, C.; Klaue, D.; Girardo, S.; Pagliara, S.; Ekpenyong, A.; Jacobi, A.; Wobus, M.; Töpfner, N.; Keyser, U. F.; Mansfeld, J.; Fischer-Friedrich, E.; Guck, J. Real-time deformability cytometry: On-the-fly cell mechanical phenotyping. Nat. Methods* ***2015,*** *12, 199–202, Copyright 2015 Nature Publishing Group. Part B: Aurich, K.; Fregin, B.; Palankar, R.; Wesche, J.; Hartwich, O.; Biedenweg, D.; Nguyen, T.-H.; Greinacher, A.; Otto, O. Label-free on chip quality assessment of cellular blood products using real-time*

(Continued)

tangentially on the cell surface and pressure acting normally deform the cell to a bullet shape. Cells are subjected to peak stress on the order of 1 kPa.[77] The theory and modeling are reported in detail by Mietke et al.[78] sDC is a popular research topic attracting much attention since Otto et al.[77] reported this single-cell mechanical phenotyping method with a real-time processing rate greater than 100 cells per second. sDC has been applied to a mass of basic cell mechanical studies,[79–81] which propels mechanical properties to be recognized as an essential factor in biology studies. Additionally, sDC has a remarkable contribution in clinical applications,[8, 82–84] yielding a high potential for sDC as a versatile tool for biology and clinical application.

Besides, sDC provides informative images of individual cell events, in addition to the deformation of cells. Toepfner et al.[8] extracted the brightness alone with size from the images to identify cell types in a blood sample. RBC can be separated by the low brightness because of the red appearance of cells and the smaller size. After the cell types are identified, the deformability is an independent parameter for assessing the functional changes. Based on sDC, Fregin et al.[85] further analyzed the dynamic cell deformation, decomposed the cell deformed shape to odd and even Fourier coefficients, which disentangle the complex cell rheology information. Furthermore, Rosendahl et al.[86] combined fluorescence detection with sDC (Fig. 4B) and presented real-time fluorescence and deformability cytometry, which connected the biochemical and biophysical properties by detecting both the cell deformation and fluorescent signal for a single-cell event. For example, in this work, the function of several proteins relating to the regulation of elastic modulus in mitotic cells is revealed. Besides, Ahsan et al.[87] integrated downstream acoustic cell sorting with fluorescence and deformability cytometry. In this work, the fluorescence detection is not only another dimension

---

Fig. 4, cont'd *deformability cytometry. Lab Chip* ***2020,*** *20, 2306-2316, Copyright 2020 Royal Society of Chemistry. Part C: Gossett, D. R.; Tse, H. T. K.; Lee, S. A.; Ying, Y.; Lindgren, A. G.; Yang, O. O.; Rao, J.; Clark, A. T.; Carlo, D. D. Hydrodynamic stretching of single cells for large population mechanical phenotyping. Proc. Natl. Acad. Sci. U. S. A.* ***2012,*** *109, 7630–7635, Copyright 2012 National Academy of Sciences. Part D: Liang, M.; Yang, D.; Zhou, Y.; Li, P.; Zhong, J.; Ai, Y. Single-cell stretching in viscoelastic fluids with electronically triggered imaging for cellular mechanical phenotyping. Anal. Chem.* ***2021,*** *93, 4567–4575, Copyright 2021 American Chemical Society. Part E: Dudani, J. S.; Gossett, D. R.; Tse, H. T. K.; Carlo, D. D. Pinched-flow hydrodynamic stretching of single-cells. Lab Chip* ***2013,*** *13, 3728–3734, Copyright 2013 Royal Society of Chemistry. Part F: Panhwar, M. H.; Czerwinski, F.; Dabbiru, V. A. S.; Komaragiri, Y.; Fregin, B.; Biedenweg, D.; Nestler, P.; Pires, R. H.; Otto, O. High-throughput cell and spheroid mechanics in virtual fluidic channels. Nat. Commun.* ***2020,*** *11, 2190, Copyright 2020 The Authors under the terms of the Creative Commons CC BY license.)*

of assessment, but also providing a molecular-based label for deep neural network training. The trained system can then sort neutrophils from RBC-depleted blood samples based on label-free cell images, obtaining a purity of 89.7%, enrichment of 10-fold. Next, the other hydrodynamic stretching method will be discussed and compared.

5.3 Extensional flow deformability cytometry

The other class in the microfluidics-based hydrodynamic stretching method is the extensional flow deformability cytometry (xDC) reported by Gossett[88] in 2012. xDC deforms cells in the center of a cross-slot, where a stagnation point is generated by the extensional flow. At this point, a cell is subjected to a normal drag force and tangential shear stress. These forces deform the cell to an ellipsoid (Fig. 4C). xDC provides high throughput of up to 2000 cells per second. Under such throughput, the recording frame rate of the high-speed camera used is 100,000 frames per second, with a minimum allowable exposure time of 1 μs to avoid image blur.[88] However, the requirement of an expensive high-speed camera and high-power illumination could be a barrier for application. To address these challenges, several works have applied viscoelastic fluids with xDC structure to stretch cells at a relatively low flow rate.[89–91]

With respect to clinical applications, Tse et al.[92] demonstrated the capability of xDC in addressing clinical problems of diagnosing malignant pleural effusions. In this complicated case, the target cells can be masked by biological noise. There is no distinct boundary between normal and abnormal cells to serve as a diagnosis reference. Therefore, a scoring system was built by machine-learning, featuring the two-dimensional distribution of deformability verse initial diameter. The trained system was then applied to quantitatively identify malignant pleural effusions with complementary results to traditional cytological studies, showing the possibility of using label-free mechanical phenotyping in intricate clinical problems.

Further, multiparameter xDC incorporating with machine learning is an interesting topic for improving classification accuracy. Lin et al.[93] extracted 21 parameters, and Masaeli et al.[94] acquired 15 parameters from the dynamic cell deformation in xDC to improve the classification accuracy. These parameters are not completely independent but suggesting the same idea that the cell size, deformability, morphology, and kinetics indicate unique cellular characteristics (morphology: the cell shape and surface roughness, kinetics: the dynamic deformability as a function of time).[94] Another way to attain multiparameter assessment is conceived by Armistead et al.[95] In this method, the deformability is investigated as a function of time. The recovery time is elicited from the function and highlighted as an important indicator that could offer more insight into cell mechanical response against the applied stress. Taken together, the images of cells passing through xDC are informative. The cellular morphology, deformation, kinetics, and recovery time can offer additional features to enhance the classification between cell populations.

Some other methods applying hydrodynamic stretching force for cellular mechanical phenotyping follow a similar idea as sDC and xDC. Deng et al.[20] employed a T junction to form the extensional flow as the cross-slot junction but achieved less horizontal drifting at the stagnation point. Because the stability of opponent flow in the symmetry cross-slot channel could be easily disrupted, the T junction could minimize the fluctuation. However, a major issue for the imaging-based hydrodynamic stretching methods is that the high recording frame rate results in a large number of invalid frames, which takes up a lot of memory space and computational power. To solve this challenge, our group integrated electrical impedance triggered, on-demand imaging and applied viscoelastic fluids to substantially reduce the invalid recording frames whilst decrease the required flow rate for sufficient cell deformation (Fig. 4D).[96] Extensional flow stretches cells, while Dudani et al.[97] reported a work that squeezes cells in the opposite direction. In this work, the flow is siphoned into a side branch and rejoined to squeeze the cell perpendicular to the flow direction (Fig. 4E), named pinched-flow hydrodynamic stretching. Then the pinched-flow and extensional flow stretching are combined and compared in the same device to employ different stress magnitudes on cells, reaping benefit from both systems. Another approach absorbed the idea of sDC and used it on the virtual fluidic channel formed by immiscible sheath flow (Fig. 4F).[98] Thus the channel dimension can be changed in the same device to measure both the single-cell and the corresponding tissues.

Comparing the sDC and xDC, both of them apply hydrodynamic stretching force under a high throughput. The essential difference comes from the magnitude of applied stress. sDC applies a smaller stress magnitude than xDC,[99] which provide options for different research interest. For example, reports pointed out that the deformation origins of xDC are dominated by nuclear changes,[20, 88] while sDC is more sensitive to cytoskeleton changes.[100] A possible reason is that the actin filament could be ruptured under high stress magnitude,[22, 91, 96] leading to an insensitivity result of cytoskeleton measuring by xDC.

6 Summary

To conclude, single-cell mechanical phenotyping is a promising label-free method for both basic science and clinical therapies. The traditional techniques offer high-precision measurement but are limited by low-throughput and skilled technician dependence. To address these challenges, microfluidics-based methods have emerged. Among them, the cDC is outstanding for low-cost, simplicity, and diversity. However, the deformability obtained from cDC entangles with factors of friction and cell size. Because of the contacting measurement way, the mechanical property interpretation is challenging. More recently, the contactless approaches sDC and xDC have arisen with merits like high throughput, reliability, repeatability, automation. Nevertheless, the requirement of a high-speed camera and high-power illumination could hamper the widespread of these methods.

Taken together, sDC is promising as a leading next-generation tool. Because sDC not only has the merits of advanced microfluidics-based methods as mentioned earlier but also has a well-developed commercialized system that has been applied in many research. What's more, the modeling and numerical study disentangles the size, deformability, and Young's Modulus. It is vital for interpreting and understanding cellular mechanical properties. And the calibration with hydrogel may take the standardization of mechanical phenotyping assessment one step further.

In fact, for decades-long, cell mechanics have not made their way as a routine application. While there are substantial convincing basic studies and applications to link cell state and disease with cell mechanical response, and attractive advantages of using a label-free marker. It could be partly due to the low-throughput limitations in conventional methods. The other mentioned problem is the lack of standardization. However, the cell mechanical phenotyping may have come to its prime. With the right tools in place, more and more researchers will join to explore the cell mechanical properties in diverse areas.

References

1. Grady, M. E.; Composto, R. J.; Eckmann, D. M. Cell elasticity with altered cytoskeletal architectures across multiple cell types. *J. Mech. Behav. Biomed. Mater.* **2016**, *61*, 197–207.
2. Pachenari, M.; Seyedpour, S. M.; Janmaleki, M.; Shayan, S. B.; Taranejoo, S.; Hosseinkhani, H. Mechanical properties of cancer cytoskeleton depend on actin filaments to microtubules content: Investigating different grades of colon cancer cell lines. *J. Biomech.* **2014**, *47*, 373–379.

3. Yamada, S.; Wirtz, D.; Kuo, S. C. Mechanics of living cells measured by laser tracking microrheology. *Biophys. J.* **2000**, *78*, 1736–1747.
4. Goddette, D. W.; Frieden, C. Actin polymerization. *J. Biol. Chem.* **1986**, *261*, 15974–15980.
5. Ribeiro, A. J. S.; Dahl, K. N. The nucleus as a central structure in defining the mechanical properties of stem cells. In Vol. 2010; *2010 Annual International Conference of the IEEE Engineering in Medicine and Biology*; 2010; pp. 831–834.
6. Guilak, F.; Tedrow, J. R.; Burgkart, R. Viscoelastic properties of the cell nucleus. *Biochem. Biophys. Res. Commun.* **2000**, *269*, 781–786.
7. Liu, H.; Wen, J.; Xiao, Y.; Liu, J.; Hopyan, S.; Radisic, M.; Simmons, C. A.; Sun, Y. In situ mechanical characterization of the cell nucleus by atomic force microscopy. *ACS Nano* **2014**, *8*, 3821–3828.
8. Toepfner, N.; Herold, C.; Otto, O.; Rosendahl, P.; Jacobi, A.; Kräter, M.; Stächele, J.; Menschner, L.; Herbig, M.; Ciuffreda, L.; Ranford-Cartwright, L.; Grzybek, M.; Coskun, Ü.; Reithuber, E.; Garriss, G.; Mellroth, P.; Henriques-Normark, B.; Tregay, N.; Suttorp, M.; Guck, J. Detection of human disease conditions by single-cell morpho-rheological phenotyping of blood. *ELife* **2018**, 7, e29213.
9. Zheng, Y.; Wen, J.; Nguyen, J.; Cachia, M. A.; Wang, C.; Sun, Y. Decreased deformability of lymphocytes in chronic lymphocytic leukemia. *Sci. Rep.* **2015**, *5*, 7613.
10. Xu, W.; Mezencev, R.; Kim, B.; Wang, L.; McDonald, J.; Sulchek, T. Cell stiffness is a biomarker of the metastatic potential of ovarian cancer cells. *PLoS One* **2012**, 7, e46609.
11. Rother, J.; Nöding, H.; Mey, I.; Janshoff, A. Atomic force microscopy-based microrheology reveals significant differences in the viscoelastic response between malign and benign cell lines. *Open Biol.* **2014**, *4*, 140046.
12. Lekka, M.; Pogoda, K.; Gostek, J.; Klymenko, O.; Prauzner-Bechcicki, S.; Wiltowska-Zuber, J.; Jaczewska, J.; Lekki, J.; Stachura, Z. Cancer cell recognition—Mechanical phenotype. *Micron* **2012**, *43*, 1259–1266.
13. Swaminathan, V.; Mythreye, K.; O'Brien, E. T.; Berchuck, A.; Blobe, G. C.; Superfine, R. Mechanical stiffness grades metastatic potential in patient tumor cells and in cancer cell lines. *Cancer Res.* **2011**, *71*, 5075–5080.
14. Shelby, J. P.; White, J.; Ganesan, K.; Rathod, P. K.; Chiu, D. T. A microfluidic model for single-cell capillary obstruction by Plasmodium falciparum-infected erythrocytes. *Proc. Natl. Acad. Sci. U. S. A.* **2003**, *100*, 14618–14622.
15. Bow, H.; Pivkin, I. V.; Diez-Silva, M.; Goldfless, S. J.; Dao, M.; Niles, J. C.; Suresh, S.; Han, J. A microfabricated deformability-based flow cytometer with application to malaria. *Lab Chip* **2011**, *11*, 1065–1073.
16. Darling, E. M.; Carlo, D. D. High-throughput assessment of cellular mechanical properties. *Annu. Rev. Biomed. Eng.* **2015**, *17*, 35–62.
17. Kozminsky, M.; Sohn, L. L. The promise of single-cell mechanophenotyping for clinical applications. *Biomicrofluidics* **2020**, *14*, 031301.
18. Xia, Y.; Wan, Y.; Hao, S.; Nisic, M.; Harouaka, R. A.; Chen, Y.; Zou, X.; Zheng, S. Nucleus of circulating tumor cell determines its translocation through biomimetic microconstrictions and its physical enrichment by microfiltration. *Small* **2018**, *14*, 1802899.
19. Rowat, A. C.; Jaalouk, D. E.; Zwerger, M.; Ung, W. L.; Eydelnant, I. A.; Olins, D. E.; Olins, A. L.; Herrmann, H.; Weitz, D. A.; Lammerding, J. Nuclear envelope composition determines the ability of neutrophil-type cells to passage through micron-scale constrictions. *J. Biol. Chem.* **2013**, *288*, 8610–8618.
20. Deng, Y.; Davis, S. P.; Yang, F.; Paulsen, K. S.; Kumar, M.; DeVaux, R. S.; Wang, X.; Conklin, D. S.; Oberai, A.; Herschkowitz, J. I.; Chung, A. J. Inertial microfluidic cell

stretcher (iMCS): Fully automated, high-throughput, and near real-time cell mechanotyping. *Small* **2017**, *13*, 1700705.
21. Lange, J. R.; Steinwachs, J.; Kolb, T.; Lautscham, L. A.; Harder, I.; Whyte, G.; Fabry, B. Microconstriction arrays for high-throughput quantitative measurements of cell mechanical properties. *Biophys. J.* **2015**, *109*, 26–34.
22. Janmey, P. A.; Euteneuer, U.; Traub, P.; Schliwa, M. Viscoelastic properties of vimentin compared with other filamentous biopolymer networks. *J. Cell Biol.* **1991**, *113*, 155–160.
23. Titushkin, I.; Cho, M. Distinct membrane mechanical properties of human mesenchymal stem cells determined using laser optical tweezers. *Biophys. J.* **2006**, *90*, 2582–2591.
24. Bennett-Guerrero, E.; Veldman, T. H.; Doctor, A.; Telen, M. J.; Ortel, T. L.; Reid, T. S.; Mulherin, M. A.; Zhu, H.; Buck, R. D.; Califf, R. M.; McMahon, T. J. Evolution of adverse changes in stored RBCs. *Proc. Natl. Acad. Sci. U. S. A.* **2007**, *104*, 17063–17068.
25. Zallen, G.; Offner, P. J.; Moore, E. E.; Blackwell, J.; Ciesla, D. J.; Gabriel, J.; Denny, C.; Silliman, C. C. Age of transfused blood is an independent risk factor for postinjury multiple organ failure. *Am. J. Surg.* **1999**, *178*, 570–572.
26. Leal-Noval, S. R.; Jara-López, I.; García-Garmendia, J. L.; Marín-Niebla, A.; Herruzo-Avilés, A.; Camacho-Laraña, P.; Loscertales, J. Influence of erythrocyte concentrate storage time on postsurgical morbidity in cardiac surgery patients. *Anesthesiology* **2003**, *98*, 815–822.
27. Weed, R. I.; LaCelle, P. L.; Merrill, E. W. Metabolic dependence of red cell deformability. *J. Clin. Investig.* **1969**, *48*, 795–809.
28. Zheng, Y.; Chen, J.; Cui, T.; Shehata, N.; Wang, C.; Sun, Y. Characterization of red blood cell deformability change during blood storage. *Lab Chip* **2013**, *14*, 577–583.
29. Skalak, R.; Branemark, P. I. Deformation of red blood cells in capillaries. *Science* **1969**, *164*, 717–719.
30. Reichel, F.; Mauer, J.; Nawaz, A. A.; Gompper, G.; Guck, J.; Fedosov, D. A. High-throughput microfluidic characterization of erythrocyte shapes and mechanical variability. *Biophys. J.* **2019**, *117*, 14–24.
31. Zeng, N. F.; Ristenpart, W. D. Mechanical response of red blood cells entering a constriction. *Biomicrofluidics* **2014**, *8*, 064123.
32. Saadat, A.; Huyke, D. A.; Oyarzun, D. I.; Escobar, P.V.; Øvreeide, I. H.; Shaqfeh, E. S. G.; Santiago, J. G. A system for the high-throughput measurement of the shear modulus distribution of human red blood cells. *Lab Chip* **2020**, *20*, 2927–2936.
33. Alibert, C.; Goud, B.; Manneville, J. Are cancer cells really softer than normal cells? *Biol. Cell* **2017**, *109*, 167–189.
34. Fletcher, D. A.; Mullins, R. D. Cell mechanics and the cytoskeleton. *Nature* **2010**, *463*, 485–492.
35. Darling, E. M.; Zauscher, S.; Block, J. A.; Guilak, F. A thin-layer model for viscoelastic, stress-relaxation testing of cells using atomic force microscopy: do cell properties reflect metastatic potential? *Biophys. J.* **2007**, *92*, 1784–1791.
36. Carlo, D. D. A mechanical biomarker of cell state in medicine. *J. Lab. Autom.* **2012**, *17*, 32–42.
37. Yamazaki, D.; Kurisu, S.; Takenawa, T. Regulation of cancer cell motility through actin reorganization. *Cancer Sci.* **2005**, *96*, 379–386.
38. Shao, J.-Y.; Ting-Beall, H. P.; Hochmuth, R. M. Static and dynamic lengths of neutrophil microvilli. *Proc. Natl. Acad. Sci. U. S. A.* **1998**, *95*, 6797–6802.
39. Shao, J. Y.; Hochmuth, R. M. Micropipette suction for measuring pico newton forces of adhesion and tether formation from neutrophil membranes. *Biophys. J.* **1996**, *71*, 2892–2901.

40. Jones, W. R.; Ting-Beall, H. P.; Lee, G. M.; Kelley, S. S.; Hochmuth, R. M.; Guilak, F. Alterations in the Young's modulus and volumetric properties of chondrocytes isolated from normal and osteoarthritic human cartilage. *J. Biomech.* **1999**, *32*, 119–127.
41. Needham, D.; Hochmuth, R. M. Rapid flow of passive neutrophils into a 4 μm pipet and measurement of cytoplasmic viscosity. *J. Biomech. Eng.* **1990**, *112*, 269–276.
42. Evans, E.; Yeung, A. Apparent viscosity and cortical tension of blood granulocytes determined by micropipet aspiration. *Biophys. J.* **1989**, *56*, 151–160.
43. Sato, M.; Theret, D. P.; Wheeler, L. T.; Ohshima, N.; Nerem, R. M. Application of the micropipette technique to the measurement of cultured porcine aortic endothelial cell viscoelastic properties. *J. Biomech. Eng.* **1990**, *112*, 263–268.
44. Rand, R. P.; Burton, A. C. Mechanical properties of the red cell membrane I. Membrane stiffness and intracellular pressure. *Biophys. J.* **1964**, *4*, 115–135.
45. Evans, E. A. New membrane concept applied to the analysis of fluid shear- and micropipette-deformed red blood cells. *Biophys. J.* **1973**, *13*, 941–954.
46. Jalili, N.; Laxminarayana, K. A review of atomic force microscopy imaging systems: Application to molecular metrology and biological sciences. *Mechatronics* **2004**, *14*, 907–945.
47. Schäpe, J.; Prauße, S.; Radmacher, M.; Stick, R. Influence of lamin A on the mechanical properties of amphibian oocyte nuclei measured by atomic force microscopy. *Biophys. J.* **2009**, *96*, 4319–4325.
48. Wu, P.-H.; Aroush, D. R.-B.; Asnacios, A.; Chen, W.-C.; Dokukin, M. E.; Doss, B. L.; Durand-Smet, P.; Ekpenyong, A.; Guck, J.; Guz, N.V.; Janmey, P.A.; Lee, J. S. H.; Moore, N. M.; Ott, A.; Poh, Y.-C.; Ros, R.; Sander, M.; Sokolov, I.; Staunton, J. R.; Wirtz, D. A comparison of methods to assess cell mechanical properties. *Nat. Meth.* **2018**, *15*, 491–498.
49. Nguyen, T. D.; Gu, Y. Investigation of cell-substrate adhesion properties of living chondrocyte by measuring adhesive shear force and detachment using AFM and inverse FEA. *Sci. Rep.* **2016**, *6*, 38059.
50. Guck, J.; Schinkinger, S.; Lincoln, B.; Wottawah, F.; Ebert, S.; Romeyke, M.; Lenz, D.; Erickson, H. M.; Ananthakrishnan, R.; Mitchell, D.; Käs, J.; Ulvick, S.; Bilby, C. Optical deformability as an inherent cell marker for testing malignant transformation and metastatic competence. *Biophys. J.* **2005**, *88*, 3689–3698.
51. Schinkinger, S.; Wottawah, F.; Travis, K.; et al. Feeling for cells with light. *Proc. SPIE* **2004**, *170*.
52. Du, E.; Dao, M.; Suresh, S. Quantitative biomechanics of healthy and diseased human red blood cells using dielectrophoresis in a microfluidic system. *Ext. Mech. Lett.* **2014**, *1*, 35–41.
53. Teng, Y.; Pang, M.; Huang, J.; Xiong, C. Mechanical characterization of cancer cells during TGF-β1-induced epithelial-mesenchymal transition using an electrodeformation-based microchip. *Sens. Act. Chem.* **2017**, *240*, 158–167.
54. Massiera, G.; Citters, K. M.V.; Biancaniello, P. L.; Crocker, J. C. Mechanics of single cells: rheology, time dependence, and fluctuations. *Biophys. J.* **2007**, *93*, 3703–3713.
55. Guck, J. Some thoughts on the future of cell mechanics. *Biophys. Rev.* **2019**, *11*, 667–670.
56. Khismatullin, D. B. The cytoskeleton and deformability of white blood cells. In Current Topics in Membranes, Vol. 64; Elsevier, 2009.
57. Byun, S.; Hecht, V. C.; Manalis, S. R. Characterizing cellular biophysical responses to stress by relating density, deformability, and size. *Biophys. J.* **2015**, *109*, 1565–1573.
58. Byun, S.; Son, S.; Amodei, D.; Cermak, N.; Shaw, J.; Kang, J. H.; Hecht, V. C.; Winslow, M. M.; Jacks, T.; Mallick, P.; Manalis, S. R. Characterizing deformability and surface friction of cancer cells. *Proc. Natl. Acad. Sci. U. S. A.* **2013**, *110*, 7580–7585.

59. Burg, T. P.; Godin, M.; Knudsen, S. M.; Shen, W.; Carlson, G.; Foster, J. S.; Babcock, K.; Manalis, S. R. Weighing of biomolecules, single cells and single nanoparticles in fluid. *Nature* **2007**, *446*, 1066–1069.
60. Fisher, T. C.; Wenby, R. B.; Meiselman, H. J. Pulse shape analysis of RBC micropore flow via new software for the cell transit analyser (CTA). *Biorheology* **1992**, *29*, 185–201.
61. Sun, T.; Morgan, H. Single-cell microfluidic impedance cytometry: A review. *Microfluid. Nanofluid.* **2010**, *8*, 423–443.
62. Chen, J.; Zheng, Y.; Tan, Q.; Shojaei-Baghini, E.; Zhang, Y. L.; Li, J.; Prasad, P.; You, L.; Wu, X. Y.; Sun, Y. Classification of cell types using a microfluidic device for mechanical and electrical measurement on single cells. *Lab Chip* **2011**, *11*, 3174–3181.
63. Zhong, J.; Yang, D.; Zhou, Y.; Liang, M.; Ai, Y. Multi-frequency single cell electrical impedance measurement for label-free cell viability analysis. *Analyst* **2021**, *146*, 1848–1858.
64. Adamo, A.; Sharei, A.; Adamo, L.; Lee, B.; Mao, S.; Jensen, K. F. Microfluidics-based assessment of cell deformability. *Anal. Chem.* **2012**, *84*, 6438–6443.
65. Zhou, Y.; Yang, D.; Zhou, Y.; Khoo, B. L.; Han, J.; Ai, Y. Characterizing deformability and electrical impedance of cancer cells in a microfluidic device. *Anal. Chem.* **2018**, *90*, 912–919.
66. Kim, J.; Han, S.; Lei, A.; Miyano, M.; Bloom, J.; Srivastava, V.; Stampfer, M. R.; Gartner, Z. J.; LaBarge, M. A.; Sohn, L. L. Characterizing cellular mechanical phenotypes with mechano-node-pore sensing. *Microsyst. Nanoeng.* **2018**, *4*, 17091.
67. Yang, D.; Zhou, Y.; Zhou, Y.; Han, J.; Ai, Y. Biophysical phenotyping of single cells using a differential multiconstriction microfluidic device with self-aligned 3D electrodes. *Biosens. Bioelectron.* **2019**, *133*, 16–23.
68. Kim, J.; Li, B.; Scheideler, O. J.; Kim, Y.; Sohn, L. L. Visco-node-pore sensing: An electronic-based microfluidic rheology platform to characterize viscoelastic properties of epithelial cells. *IScience* **2019**, *13*, 214–228.
69. Nyberg, K. D.; Hu, K. H.; Kleinman, S. H.; Khismatullin, D. B.; Butte, M. J.; Rowat, A. C. Quantitative deformability cytometry: Rapid, calibrated measurements of cell mechanical properties. *Biophys. J.* **2017**, *113*, 1574–1584.
70. Lee, L. M.; Liu, A. P. A microfluidic pipette array for mechanophenotyping of cancer cells and mechanical gating of mechanosensitive channels. *Lab Chip* **2015**, *15*, 264–273.
71. Rosenbluth, M. J.; Lam, W. A.; Fletcher, D. A. Analyzing cell mechanics in hematologic diseases with microfluidic biophysical flow cytometry. *Lab Chip* **2008**, *8*, 1062–1070.
72. Davidson, P. M.; Fedorchak, G. R.; Mondésert-Deveraux, S.; Bell, E. S.; Isermann, P.; Aubry, D.; Allena, R.; Lammerding, J. High-throughput microfluidic micropipette aspiration device to probe time-scale dependent nuclear mechanics in intact cells. *Lab Chip* **2019**, *19*, 3652–3663.
73. Chen, Z.; Zhu, Y.; Xu, D.; Alam, M. M.; Shui, L.; Chen, H. Cell elasticity measurement using a microfluidic device with real-time pressure feedback. *Lab Chip* **2020**, *20*, 2343–2353.
74. Hou, H. W.; Li, Q. S.; Lee, G. Y. H.; Kumar, A. P.; Ong, C. N.; Lim, C. T. Deformability study of breast cancer cells using microfluidics. *Biomed. Microdev.* **2009**, *11*, 557–564.
75. Nyberg, K. D.; Scott, M. B.; Bruce, S. L.; Gopinath, A. B.; Bikos, D.; Mason, T. G.; Kim, J. W.; Choi, H. S.; Rowat, A. C. The physical origins of transit time measurements for rapid, single cell mechanotyping. *Lab Chip* **2016**, *16*, 3330–3339.
76. Raj, A.; Dixit, M.; Doble, M.; Sen, A. K. A combined experimental and theoretical approach towards mechanophenotyping of biological cells using a constricted microchannel. *Lab Chip* **2017**, *17*, 3704–3716.

77. Otto, O.; Rosendahl, P.; Mietke, A.; Golfier, S.; Herold, C.; Klaue, D.; Girardo, S.; Pagliara, S.; Ekpenyong, A.; Jacobi, A.; Wobus, M.; Töpfner, N.; Keyser, U. F.; Mansfeld, J.; Fischer-Friedrich, E.; Guck, J. Real-time deformability cytometry: On-the-fly cell mechanical phenotyping. *Nat. Methods* **2015**, *12*, 199–202.
78. Mietke, A.; Otto, O.; Girardo, S.; Rosendahl, P.; Taubenberger, A.; Golfier, S.; Ulbricht, E.; Aland, S.; Guck, J.; Fischer-Friedrich, E. Extracting cell stiffness from real-time deformability cytometry: Theory and experiment. *Biophys. J.* **2015**, *2023–2036*, 109.
79. Urbanska, M.; Winzi, M.; Neumann, K.; Abuhattum, S.; Rosendahl, P.; Müller, P.; Taubenberger, A.; Anastassiadis, K.; Guck, J. Single-cell mechanical phenotype is an intrinsic marker of reprogramming and differentiation along the mouse neural lineage. *Development* **2017**, *144*, 4313–4321.
80. Tavares, S.; Vieira, A. F.; Taubenberger, A. V.; Araújo, M.; Martins, N. P.; Brás-Pereira, C.; Polónia, A.; Herbig, M.; Barreto, C.; Otto, O.; Cardoso, J.; Pereira-Leal, J. B.; Guck, J.; Paredes, J.; Janody, F. Actin stress fiber organization promotes cell stiffening and proliferation of pre-invasive breast cancer cells. *Nat. Commun.* **2017**, *8*, 15237.
81. Munder, M. C.; Midtvedt, D.; Franzmann, T.; Nüske, E.; Otto, O.; Herbig, M.; Ulbricht, E.; Müller, P.; Taubenberger, A.; Maharana, S.; Malinovska, L.; Richter, D.; Guck, J.; Zaburdaev, V.; Alberti, S. A pH-driven transition of the cytoplasm from a fluid- to a solid-like state promotes entry into dormancy. *ELife* **2016**, *5*, e09347.
82. Xavier, M.; Rosendahl, P.; Herbig, M.; Kräter, M.; Spencer, D.; Bornhäuser, M.; Oreffo, R. O. C.; Morgan, H.; Guck, J.; Otto, O. Mechanical phenotyping of primary human skeletal stem cells in heterogeneous populations by real-time deformability cytometry. *Integr. Biol.* **2016**, *8*, 616–623.
83. Guzniczak, E.; Zadeh, M. M.; Dempsey, F.; Jimenez, M.; Bock, H.; Whyte, G.; Willoughby, N.; Bridle, H. High-throughput assessment of mechanical properties of stem cell derived red blood cells, toward cellular downstream processing. *Sci. Rep.* **2017**, 7, 14457.
84. Aurich, K.; Fregin, B.; Palankar, R.; Wesche, J.; Hartwich, O.; Biedenweg, D.; Nguyen, T.-H.; Greinacher, A.; Otto, O. Label-free on chip quality assessment of cellular blood products using real-time deformability cytometry. *Lab Chip* **2020**, *20*, 2306–2316.
85. Fregin, B.; Czerwinski, F.; Biedenweg, D.; Girardo, S.; Gross, S.; Aurich, K.; Otto, O. High-throughput single-cell rheology in complex samples by dynamic real-time deformability cytometry. *Nat. Commun.* **2019**, *10*, 415.
86. Rosendahl, P.; Plak, K.; Jacobi, A.; Kraeter, M.; Toepfner, N.; Otto, O.; Herold, C.; Winzi, M.; Herbig, M.; Ge, Y.; Girardo, S.; Wagner, K.; Baum, B.; Guck, J. Real-time fluorescence and deformability cytometry. *Nat. Meth.* **2018**, *15*, 355–358.
87. Nawaz, A. A.; Urbanska, M.; Herbig, M.; Nötzel, M.; Kräter, M.; Rosendahl, P.; Herold, C.; Toepfner, N.; Kubánková, M.; Goswami, R.; Abuhattum, S.; Reichel, F.; Müller, P.; Taubenberger, A.; Girardo, S.; Jacobi, A.; Guck, J. Intelligent image-based deformation-assisted cell sorting with molecular specificity. *Nat. Meth.* **2020**, *17*, 595–599.
88. Gossett, D. R.; Tse, H. T. K.; Lee, S. A.; Ying, Y.; Lindgren, A. G.; Yang, O. O.; Rao, J.; Clark, A. T.; Carlo, D. D. Hydrodynamic stretching of single cells for large population mechanical phenotyping. *Proc. Natl. Acad. Sci. U. S. A.* **2012**, *109*, 7630–7635.
89. Cha, S.; Shin, T.; Lee, S. S.; Shim, W.; Lee, G.; Lee, S. J.; Kim, Y.; Kim, J. M. Cell stretching measurement utilizing viscoelastic particle focusing. *Anal. Chem.* **2012**, *84*, 10471–10477.
90. Armistead, F. J.; Pablo, J. G. D.; Gadêlha, H.; Peyman, S. A.; Evans, S. D. Cells under stress: An inertial-shear microfluidic determination of cell behaviour. *Biophys. J.* **2019**, *116*, 1127–1135.
91. Guillou, L.; Dahl, J. B.; Lin, J.-M. G.; Barakat, A. I.; Husson, J.; Muller, S. J.; Kumar, S. Measuring cell viscoelastic properties using a microfluidic extensional flow device. *Biophys. J.* **2016**, *111*, 2039–2050.

92. Tse, H. T. K.; Gossett, D. R.; Moon, Y. S.; Masaeli, M.; Sohsman, M.; Ying, Y.; Mislick, K.; Adams, R. P.; Rao, J.; Carlo, D. D. Quantitative diagnosis of malignant pleural effusions by single-cell mechanophenotyping. *Sci. Transl. Med.* **2013**, *5*. 212ra163-212ra163.
93. Lin, J.; Kim, D.; Tse, H. T.; Tseng, P.; Peng, L.; Dhar, M.; Karumbayaram, S.; Carlo, D. D. High-throughput physical phenotyping of cell differentiation. *Microsyst. Nanoeng.* **2017**, *3*, 17013.
94. Masaeli, M.; Gupta, D.; O'Byrne, S.; Tse, H. T. K.; Gossett, D. R.; Tseng, P.; Utada, A. S.; Jung, H.-J.; Young, S.; Clark, A. T.; Carlo, D. D. Multiparameter mechanical and morphometric screening of cells. *Sci. Rep.* **2016**, *6*, 37863.
95. Armistead, F. J.; Pablo, J. G. D.; Gadêlha, H.; Peyman, S. A.; Evans, S. D. Physical biomarkers of disease progression: On-chip monitoring of changes in mechanobiology of colorectal cancer cells. *Sci. Rep.* **2020**, *10*, 3254.
96. Liang, M.; Yang, D.; Zhou, Y.; Li, P.; Zhong, J.; Ai, Y. Single-cell stretching in viscoelastic fluids with electronically triggered imaging for cellular mechanical phenotyping. *Anal. Chem.* **2021**, *93*, 4567–4575.
97. Dudani, J. S.; Gossett, D. R.; Tse, H. T. K.; Carlo, D. D. Pinched-flow hydrodynamic stretching of single-cells. *Lab Chip* **2013**, *13*, 3728–3734.
98. Panhwar, M. H.; Czerwinski, F.; Dabbiru, V. A. S.; Komaragiri, Y.; Fregin, B.; Biedenweg, D.; Nestler, P.; Pires, R. H.; Otto, O. High-throughput cell and spheroid mechanics in virtual fluidic channels. *Nat. Commun.* **2020**, *11*, 2190.
99. Urbanska, M.; Muñoz, H. E.; Bagnall, J. S.; Otto, O.; Manalis, S. R.; Carlo, D. D.; Guck, J. A comparison of microfluidic methods for high-throughput cell deformability measurements. *Nat. Meth.* **2020**, *17*, 587–593.
100. Golfier, S.; Rosendahl, P.; Mietke, A.; Herbig, M.; Guck, J.; Otto, O. High-throughput cell mechanical phenotyping for label-free titration assays of cytoskeletal modifications. *Cytoskeleton* **2017**, *74*, 283–296.

CHAPTER 6

Single-cell analysis by evanescent wave sensing and hyperspectral microscopy

Ziwei Liu[a,b], Jian Chen[a], and Zhi-Mei Qi[a,b,*]
[a]State Key Laboratory of Transducer Technology, Aerospace Information Research Institute, Chinese Academy of Sciences, Beijing, China
[b]University of Chinese Academy of Sciences, Beijing, China

1 Introduction

Single-cell analysis (SCA) is an advanced bio-analytical technology involving physics, chemistry, biology, biomedicine, life science, materials science, nanotechnology, computer statistics, and so on, which takes individual cells as the object of investigation to rapidly obtain their physical, chemical, biological, and structural properties. SCA has been widely used for cell counting, cell sorting, cell-targeted therapy, early cancer diagnosis, study of intercellular, and intracellular interactions, etc., and now it has become the main tool for cell biology and biomedical research. The birth and rapid development of SCA is closely dependent on the development of various advanced optical technologies, including laser technology, optical sensing technology, optical microscopy, and CCD imager technology. The optical biosensing is performed by measuring the changes of wavelength, amplitude, phase, polarization or photon lifetime of the output light, and these changes result from the interaction between optical field and biological analytes. Signal of an optical biosensor carrying the quantitative information of biological analytes is generated through resonance, fluorescence, scattering, absorption, interference, and other optical mechanisms, which is detected with either a silicon photodetector or a CCD imaging sensor or a fiber-optic spectrometer. Quantitative information such as analyte concentration and kinetic parameters, can be extracted from optical signals. Nowadays a large variety of optical biosensors are available, including optical fiber biosensors, interferometric biosensors, photonic crystal sensors, grating-coupled

*Corresponding author.

Biosensors for Single-Cell Analysis
https://doi.org/10.1016/B978-0-323-89841-6.00003-7

waveguide resonance (WGR) biosensors, surface plasmon resonance (SPR) biosensors, metal-cladded plasmon waveguide resonance (PWR) biosensors, surface-enhanced Raman scattering (SERS) sensors, etc. Among them, SPR-based sensors are one of the most reported and most widely used optical biosensors, which have been successfully applied in the fields ranging from single-molecule sensing to single-cell imaging.[1] SERS-based optical biosensors with molecular fingerprint detection capability are well employed in the study of single-cell photo physiology, such as dynamic monitoring of cell metabolism and cellular activities.[2, 3] Regarding the optical transduction mechanisms and their sensing characteristics of the above-mentioned optical biosensors to non-cell analytes, we invite readers to refer to the reported work,[4, 5] which describes different types of optical biosensors in detail.

In this chapter we mainly review the recent applications of optical biosensors based on evanescent-wave sensing and advanced spectral imaging technology at a single-cell level. Optical biosensors utilize evanescent field to interact with measurands at the surface and are very sensitive to the change of surface refractive index. SPR biosensors which use the *p*-polarized light to excite surface plasmon wave at the metal/dielectric interface, and WGR biosensors which use diffraction of nanograting waveguide to generate an evanescent wave are two label-free refractometric biosensors successfully applied in cell biology.[6] Each cell component with a distinct refractive index value is easily detected by those refractometric biosensors. In addition, the high spatial resolution required for cell analysis can be achieved through microscopy, and some molecular level details can be provided through spectroscopy, which has greatly promoted the development of cell analysis methods.[7] The emerging spectral imaging methods open ample avenues for scientists to carry out more in-depth research on single cells. As an example, hyperspectral imaging (HSI) technology combines advanced spectroscopy and imaging techniques, having the advantage of providing both spatial and spectral information about an object. Differences in light scattering and absorption from cells, which are mainly dependent on the refractive index, cell thickness, and molecular composition of subcellular level organelles and cellular walls, can be captured by a HSI system with high spatial and spectral resolution.[8] Therefore, herein we aim to introduce the progress on label-free optical refractometric biosensors and the development of HSI-based systems for characterization of individual cells.

2 Label-free optical refractometric biosensors for SCA

2.1 Surface plasmon resonance microscopy and imaging methods

SPR phenomenon results from oscillation of free electron density induced by incident light at the metal/dielectric interface, initially observed by Wood in 1902.[9] Since the first successful biosensing application in 1983,[10] SPR sensors have attracted attentions in a broad scope of applications. In 1987, Yeatman et al.[11] prepared the first monochromatic SPR imaging sensor using a He-Ne laser light source based on Kretschmann configuration. Combined with CCD cameras and microscopes which can provide spatial resolution, SPR imaging and SPR microscopy (SPRM) enable for single-cell level analysis. As shown in Fig. 1, three typical SPR-based techniques are presented. Conventional SPR devices with photodiodes are usually used to monitor the kinetics of biomolecular interactions, enabling only for single-point averaged measurements with temporal resolution.

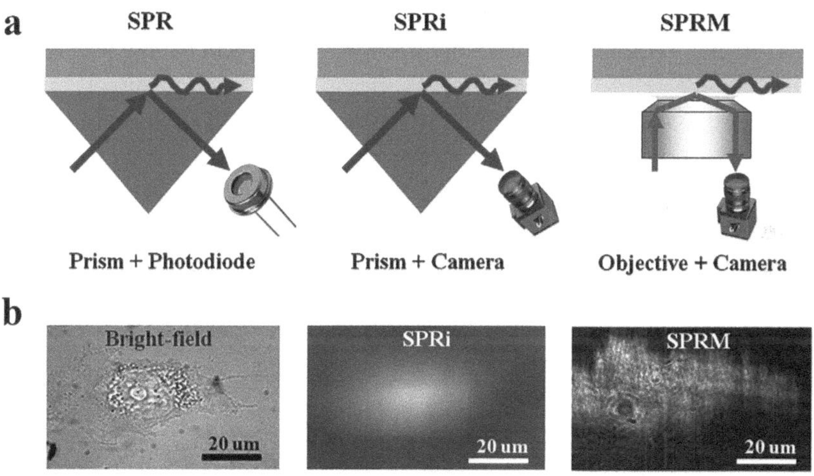

Fig. 1 (A) Three types of surface plasmon resonance (SPR) setups invented in the past three decades: from single-point kinetic measurement with only temporal resolution to two-dimensional imaging with both spatial and temporal resolutions; (B) *Left*: a bright-field image of a single cell acquired by normal optical microscope; *Middle*: an SPR image with onefold to fivefold magnification acquired by prism-based setups for the same cell; *Right*: a SPR image with 60- to 100-fold magnification acquired by objective-based setups for the same cell. *Reproduced with permission Su, Y. W.; Wang, W. Surface plasmon resonance sensing: From purified biomolecules to intact cells. Anal. Bioanal. Chem.* ***2018,*** *410, 3943–3951. Copyright 2018, Springer-Verlag GmbH Germany, part of Springer Nature.*

Combined with CCD cameras, SPR setups can carry out two-dimensional studies of living cells and microarrays. Go a step further, replacing the prism with a high numerical aperture objective is able to improve the spatial resolution up to the optical diffraction limit without image distortion,[12–14] which is advantageous for the cellular and subcellular study.

As an evanescent-wave sensing technique, SPR sensor is sensitive to minute changes of refractive index at surface/interface. When cells interfere with surface evanescent field, the propagation constant of surface plasmon wave will be modified, and the intensity of the reflected light or the resonance angle is determined by cell–substrate distances, which can be used to study cell/substrate contacts. Early in 1999, Giebel et al.[15] first built a SPRM to analyze the contacts between a primary goldfish glial cell and a glass substrate coated by aluminum film for excitation of surface plasmon wave, as a result, the membrane/substrate distances in the nanometer scale were visualized and quantified. The cultured cell acts as a dielectric fixed on the surface of SPR biosensors, and the cell–substrate distances of different sites are quantified by calculating variations of the resonance angle and visualized by CCD capturing the difference of reflected light intensities. Wang et al.[16] first attempted to use SPRM with subcellular spatial resolution to study cell–substrate adhesion strength when cells were simulated by extracellular osmotic pressure. As shown in Fig.2, individual cells with hypertonic stimulation shrank and detached from the substrate, creating a vertical displacement to the substrate. SPR biosensors were very sensitive to the refractive index changes caused by distance variation between the cell membrane and the substrate; therefore, quantitative local adhesion information can be exacted from the SPR signals.

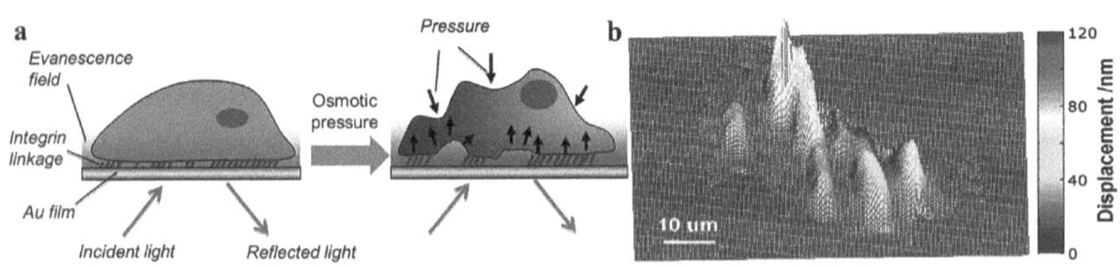

Fig. 2 (A) Schematic illustration of mapping with cell–substrate interactions with SPRM and (B) mapping the local cell–substrate adhesion strength, where a larger displacement reflects a weaker adhesion strength. *Reproduced with permission Wang, W.; Wang, S.; Liu, Q.; Wu, J.; Tao, N. Mapping single-cell-substrate interactions by surface plasmon resonance microscopy. Langmuir* ***2012,*** *28, 13373–13379. Copyright 2012, American Chemical Society.*

To improve the spatial resolution of SPRM system, Toma et al.[17] employed a lens-imaging-type SPM (LISPM) to observe the whole morphological characteristics of human embryonic kidney 293 cells in real-time and a scanning-localized SPM (SLSPM) to quantify cleft gap distance (shown in Fig.3). The cellular adhesion differences in various peptides or proteins coated SPR biosensors were analyzed by mapping the inhomogeneous distribution of cell–electrode cleft gap distance. The gap distance was determined by the eighth-degree polynomial fitting function between the distance and SPR propagation constant. Kreysing et al.[18] also used the above two mentioned SPRM systems to study the dynamics of beating cardiomyocyte. As shown in Fig.4, the cell–substrate distance of contracting cardiomyocyte and the cytosolic refractive index variations were measured to analyze the cellular response process to extracellular stimulus. Nanometer localization precision and millisecond temporal resolution are necessary for fast intracellular dynamic processes to understand the cellular behaviors at the molecular scale.[19] Utilizing the surface plasmonic waves scattered by single organelles, high contrast images can be obtained to track organelle dynamics in single cells.

To expand the applications of SPRM in research of single living cells, researchers have explored versatile SPRM systems integrated with other optical techniques. For example, Wang et al.[14] designed a microscope system that allowed for simultaneous bright-field, fluorescence, and SPRM imaging for a single cell (shown in Fig.5). In both label-free and label-based modes, distribution and local binding kinetics of membrane proteins in the cell membrane were mapped and quantified. Different from fluorescence imaging the whole cell membrane, SPRM images focused on the membrane proteins within the evanescent field. Lu and Liu groups[20, 21] developed a plasmonic-based electrochemical impedance microscope (P-EIM) with excellent spatiotemporal resolution by combining a high-resolution SPRM with alternating current modulation. As shown in Fig. 6, that P-EIM system was able to monitor local cell-signaling processes and propagation of action potential within single neurons, which is helpful to understand various cellular events and behaviors of single cells.

Earlier-mentioned SPRM systems are constructed based on intensity interrogation using the monochromatic light source, which is fast and effective but suffers from limited dynamic range. Specific incident angle and incident light at a certain excited wavelength are required and optimized for best possible sensitivity, which is usually not practical to obtain the optimal response at one operating point for simultaneous detection

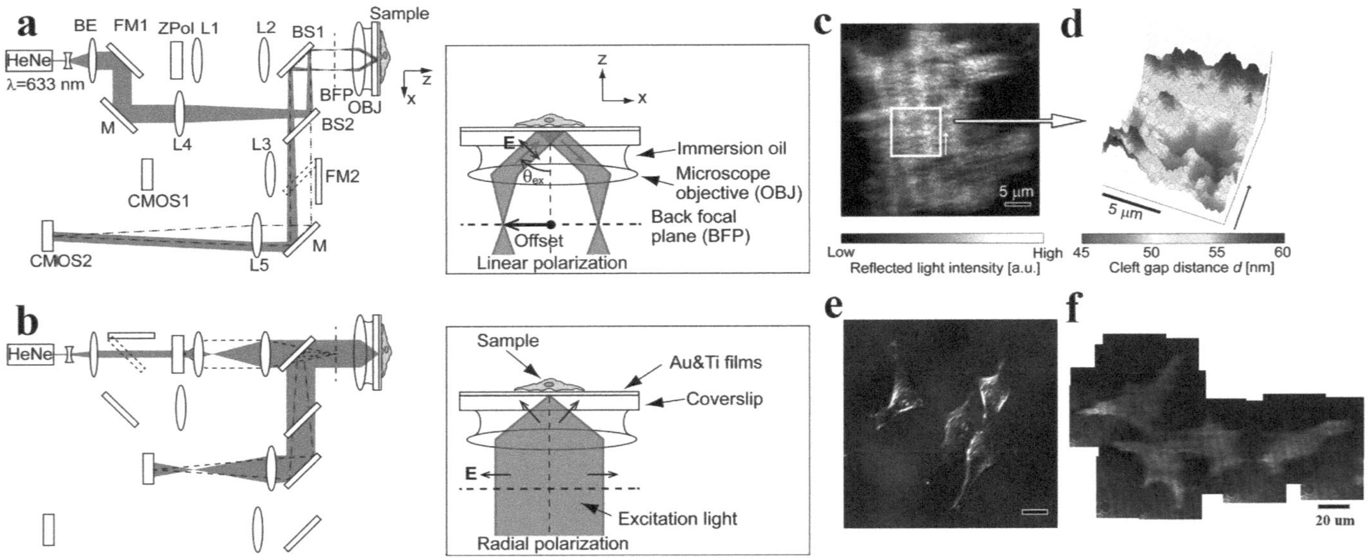

Fig. 3 Schematics of (A) LISPM and (B) SLSPM; representative (C) LISPM and (D) SLSPM images of 293 cells grown on poly-L-lysine-coated gold surface; Comparison of (E) differential interference contrast microscope and (F) LISPM images. *Adapted with permission Toma, K.; Kano, H.; Offenhaeusser, A. Label-free measurement of cell-electrode cleft gap distance with high spatial resolution surface plasmon microscopy. ACS Nano* ***2014,*** *8, 12612–12619. Copyright 2014, American Chemical Society.*

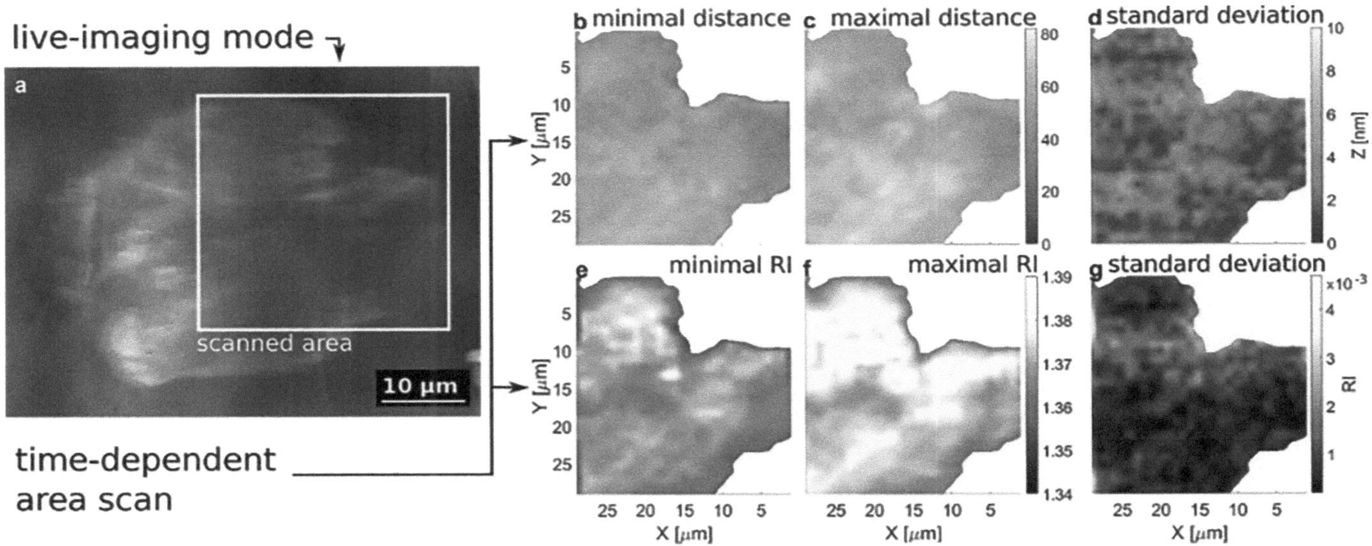

Fig. 4 (A) The live-imaging mode was used to define the scanning area (*white frame*) at the cell–substrate interface. (B) The minimal distance corresponds to the contracted cell while the maximal distance (C) corresponds to the resting state. (D) The standard deviation indicates the membrane dynamics. The refractive index of the cytosol varies between the (E) minimal and (F) maximal values with the (G) standard deviation. *Reproduced with permission Kreysing, E.; Hassani, H.; Hampe, N.; Offenhaeusser, A. Nanometer-resolved mapping of cell–substrate distances of contracting cardiomyocytes using surface plasmon resonance microscopy. ACS Nano* ***2018,*** *12, 8934–8942. Copyright 2018, American Chemical Society.*

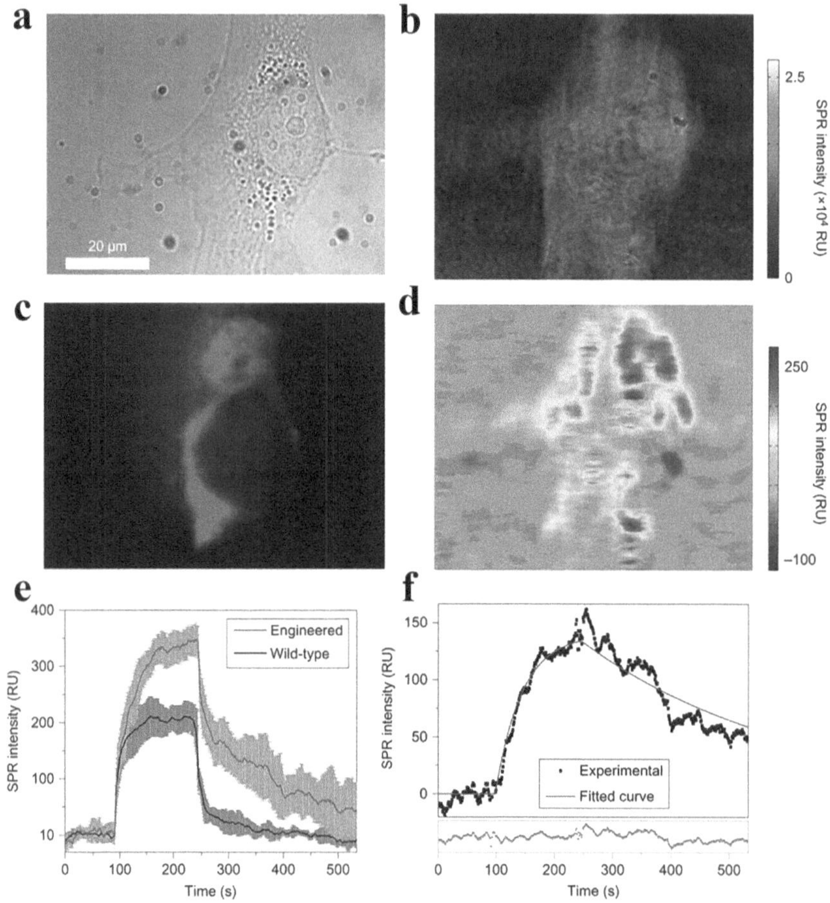

Fig. 5 (A) Bright-field and (B) SPRM images of an engineered SH-EP1 human epithelial cells that expressed nicotinic acetylcholine receptors (nAChRs). (C) The immunofluorescence image of the same cell obtained by sequentially incubating with primary and Alexa Fluor 488-labeled secondary antibody. (D) Distribution of nAChRs of the same cell resolved by the binding of primary antibody to the cell. (E) SPR sensorgrams of engineered (*red curve, black in the print version*) and wild-type (*black curve*) SH-EP1 cells during primary-antibody binding. Five cells were examined in each case and variation was shown by the *pink* (*light gray in the print version*) and *gray curves*. (F) The sensorgram (*black dots*) after correction for the bulk refractive index effect (by taking the wild-type as reference), where the *red curve* (*black in the print version*) was fitted to the first-order kinetics model. The fitting residuals were shown in the lower portion of the figure. *Reproduced with permission Wang, W.; Yang, Y. Z.; Wang, S. P.; Nagaraj, V. J.; Liu, Q.; Wu, J., et al. Label-free measuring and mapping of binding kinetics of membrane proteins in single living cells. Nat. Chem.* ***2012,*** *4, 846–853. Copyright 2012, Nature Publishing Group.*

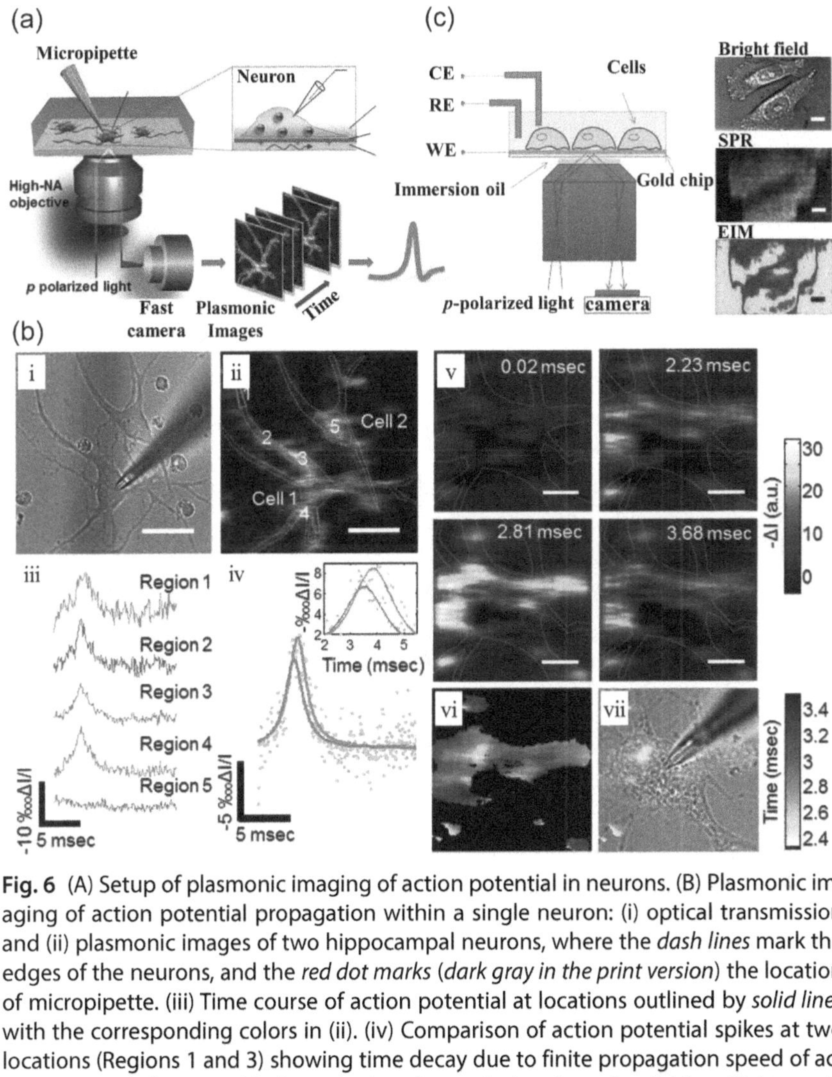

Fig. 6 (A) Setup of plasmonic imaging of action potential in neurons. (B) Plasmonic imaging of action potential propagation within a single neuron: (i) optical transmission and (ii) plasmonic images of two hippocampal neurons, where the *dash lines* mark the edges of the neurons, and the *red dot marks* (*dark gray in the print version*) the location of micropipette. (iii) Time course of action potential at locations outlined by *solid lines* with the corresponding colors in (ii). (iv) Comparison of action potential spikes at two locations (Regions 1 and 3) showing time decay due to finite propagation speed of action potential; the *green* (*dark gray in the print version*) and *purple solid lines* (*light gray in the print version*) are Lorentzian fitting to the corresponding P-EIM data. (v) Snapshots of action potential propagation at a frame rate of 100,000 fps from a neuron. (vi) Action potential delay map. (vii) Overlay of the action potential delay map with bright field image of the neuron; Scale bar = 20 μm. (C) *Left*: setup of the plasmonic-based electrochemical impedance microscope (P-EIM); *CE*, counter electrode; *NA*, numeric aperture; *RE*, reference electrode; *WE*, working electrode; *Right*: bright-field, SPR, and EIM images of HeLa cells; Scale bar = 10 μm. *Part A and B: Adapted with permission Liu, X. W.; Yang, Y.; Wang, W.; Wang, S.; Gao, M.; Wu, J. et al. Plasmonic-based electrochemical impedance imaging of electrical activities in single cells. Angew. Chem. Int. Ed. Engl.* ***2017,*** *56, 8855–8859. Copyright 2017, Wiley-VCH Verlag GmbH & Co. KGaA, Weinheim. Part C: Adapted with permission Lu, J.; Li, J. Label-free imaging of dynamic and transient calcium signaling in single cells. Angew. Chem. Int. Ed. Engl.* ***2015,*** *54, 13576–13580. Copyright 2015, WILEY-VCH Verlag GmbH & Co. KGaA, Weinheim.*

of multiple interactions.[22, 23] In 1996, Knobloch et al.[24] prepared the first colorful SPR imaging sensor using a white light source. Compared with the monochromatic SPR imaging sensor, the colorful SPR image has a large amount of information and a wide dynamic range. A resonance color image can use different colors to reflect the biochemical reactions of different areas, so that one imaging can cover the biochemical of all sites in the entire imaging area. Relatively, wavelength-scanning SPRM (WS-SPRM) systems, which allow for operating in wavelength interrogation, are flexible to achieve high-throughput, real-time, and high-sensitivity biosensing.[25, 26] Recently, Zeng et al.[27] first proposed a novel method based on a WS-SPRM to detect cell–substrate interactions (shown in Fig. 7), which showed a larger dynamic range than that of intensity interrogation-based SPRM. Moreover, in the future, the developed multispectral SPR imaging equipment using a liquid crystal tunable filter (LCTF) or an acousto-optic tunable filter (AOTF),[28, 29] and the hyperspectral SPR imaging system using a hyperspectral imager will become powerful tools for high-throughput and high-resolution analysis of single cells.[30, 31]

2.2 Optical waveguides biosensors

Optical waveguides biosensors have attracted increasing attention in the field of cellular assays, which offer a simple label free way to study cells adhesion on various surfaces as the guided mode sensitive to the refractive index changes near waveguide surface. The first application of optical waveguides to cells is to determine the number and size of cells attached to the substrate.[32] Evanescent-wave probing depth of conventional waveguides with a high refractive index has an upper limit of 100–150 nm, which is low-sensitive to the cell. Thus the alternative reverse symmetry waveguide sensors in which the waveguide layer has a lower refractive index than the covered layer are developed for providing a deeper probing depth, successfully to investigate the attachment and spreading of living cells (shown in Fig. 8A).[33–35] In the studying of cellular responses, WGR biosensors (Fig. 8B) have compelling advantages to measure dynamic mass redistribution (DMR) of cellular contents in both perpendicular and parallel directions to the sensor surface (Fig. 8C), which is achieved by analyzing changes of the incident angle and the shape of resonant peaks.[36–38] Moreover, a microfluidic WGR biosensor system shown in Fig. 8D expands the application of DMR assays for cellular signaling (Fig. 8E).[39] Label-free WGR biosensors also make it possible to noninvasively and continuously monitor cancer signaling and cellular processes.[40, 41] Integration of WGR biosensors with fluorescence

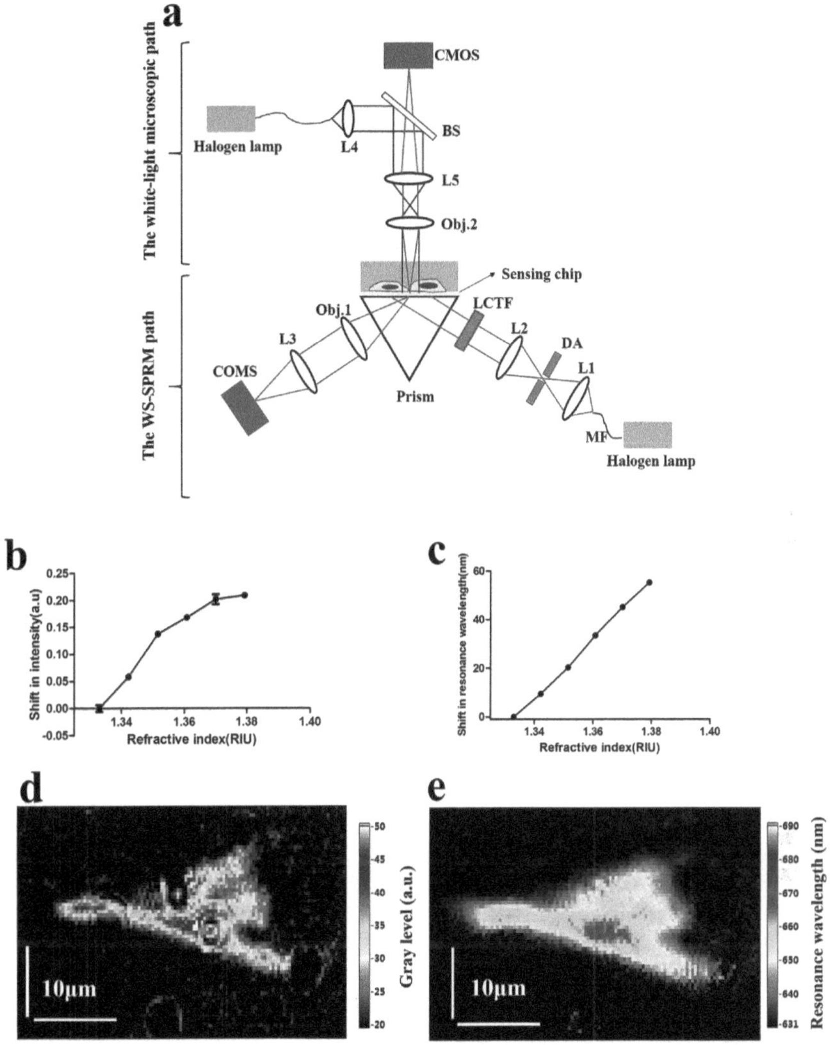

Fig. 7 (A) The schematic of the WS-SPRM setup. (B) The relationship between the intensity shift and the increasing refractive index. (C) The relationship between the resonance dip shift with the increasing refractive index and each point represents the mean ± SD of nine regions of interest. (D) The intensity image of a single cell acquired by the I-SPRM. (E) The resonance wavelength image of a single cell acquired by the WS-SPRM. *Reproduced with permission Zeng, Y. J.; Zhou, J.; Wang, X. L.; Cai, Z. W.; Shao, Y. H. Wavelength-scanning surface plasmon resonance microscopy:a novel tool for real time sensing of cell-substrate interactions. Biosens. Bioelectron.* ***2019,*** *145, 111717. Copyright 2019, Elsevier B.V.*

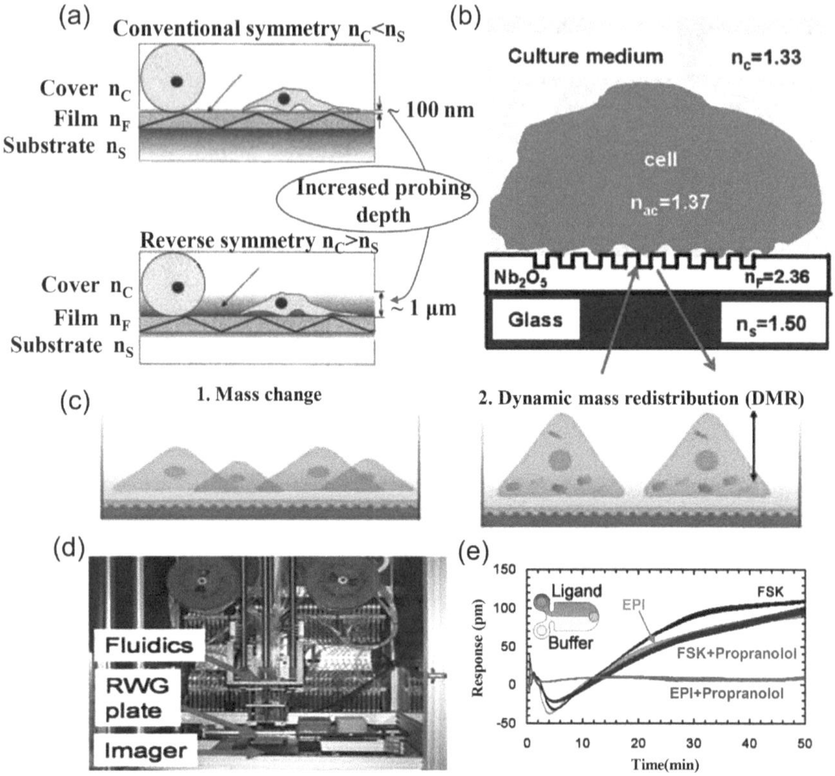

Fig. 8 (A) Conventional and reverse symmetry waveguide design. (B) Schematic drawing of the WGR biosensor for sensing living cells. (C) Two main cellular responses: (1) mass change and (2) DMR. (D) A microfluidic WGR biosensor system for whole cell sensing. (E) The DMR of cells responding to perfusion with different solutions: 5 nM epinephrine (EPI) in the absence and presence of 1 μM propranolol, and 5 μM forskolin (FSK) in the absence and presence of 1 μM propranolol. *Part A: Adapted with permission Horvath, R.; Pedersen, H. C.; Skivesen, N.; Selmeczi, D.; Larsen, N. B. Monitoring of living cell attachment and spreading using reverse symmetry waveguide sensing. Appl. Phys. Lett.* ***2005,*** *86, 071101. Copyright 2005, AIP Publishing. Part B: Adapted with permission Fang, Y.; Ferrie, A. M.; Fontaine, N. H.; Yuen, P. K. Characteristics of dynamic mass redistribution of epidermal growth factor receptor signaling in living cells measured with label-free optical biosensors. Anal. Chem.* ***2005,*** *77, 5720–5725. Copyright 2005, American Chemical Society. Part C: Adapted with permission Nazirizadeh, Y.; Behrends, V.; Prosz, A.; Orgovan, N.; Horvath, R.; Ferrie, A. M., et al. Intensity interrogation near cutoff resonance for label-free cellular profiling. Sci. Rep.* ***2016,*** *6, 24685. Part D and E: Adapted with permission Ferrie, A. M.; Wu, Q.; Fang, Y. Microfluidic resonant waveguide grating biosensor system for whole cell sensing. Appl. Phys. Lett.* ***2011,*** *98, 163703. Copyright 2011, AIP Publishing.*

microscopy provides a dual mode to characterize the receptor signaling that interfere with cell adhesion,[42] where the green fluorescent protein (GFP) can be excited by the evanescent wave propagating along the WGR diffractive grating interface. In addition to fluorescent approach, combined with robotic fluidic force microscopy,[43] WGR-based optical biosensors can be used to monitor the adhesion of individual cancer cells with high spatial and temporal resolution, which opens up a new perspective for evaluating the adhesion kinetics of single cells.

When a dielectric waveguide layer is coated on the conventional SPR biosensing chip, the so-called plasmon-waveguide resonance (PWR) biosensor is formed,[44–48] which can be used to sensitively detect the cell–dielectric interaction rather than cell–metal interaction. Compared to SPR sensing, PWR biosensors have deeper probing depth able for detecting the microscopic targets such as cells.[49] A PWR-based microscopy system with the ability of resolving single cells is developed to label-free investigate and quantify cell-signaling activity induced by external stimuli (shown in Fig.9).[50] The system is capable of high-contrast selective imaging of the cell/matrix interface, and thus making it an important advantage in studying the heterogeneity of cellular responses in small cell populations. Furthermore, combined with surface enhanced fluorescence technique,[51] the PWR microscopy system owes the capability of providing both label-free imaging and label-based signals of receptor-dependent cellular signaling activity, which has promising applications in the field of cell biology and pharmacology.

3 Hyperspectral microscopic imaging methods for SCA

HSI is a non-invasive spectral imaging technique initially applied for satellite remote sensing. Different from conventional RGB images, HSI data are stored in the form of a data hypercube that contains detailed spectral signature for each pixel in an image, as shown in Fig. 10, and the spectral range can cover from ultraviolet to near-infrared (200–2500 nm). The application of HSI technology in the macro field involves astronomy, agriculture, mining and geology, food quality and safety control, medicine, etc. In recent years, multi-modal HSI systems are successfully developed by coupling a hyperspectral imager to various microscopes (dark-filed, confocal or fluorescence microscopes) and/or traditionally single point spectroscopy (fluorescence, Fourier-transform infrared, near infrared, Raman spectroscopy, and so on). In those systems, HSI can be carried out in reflectance,

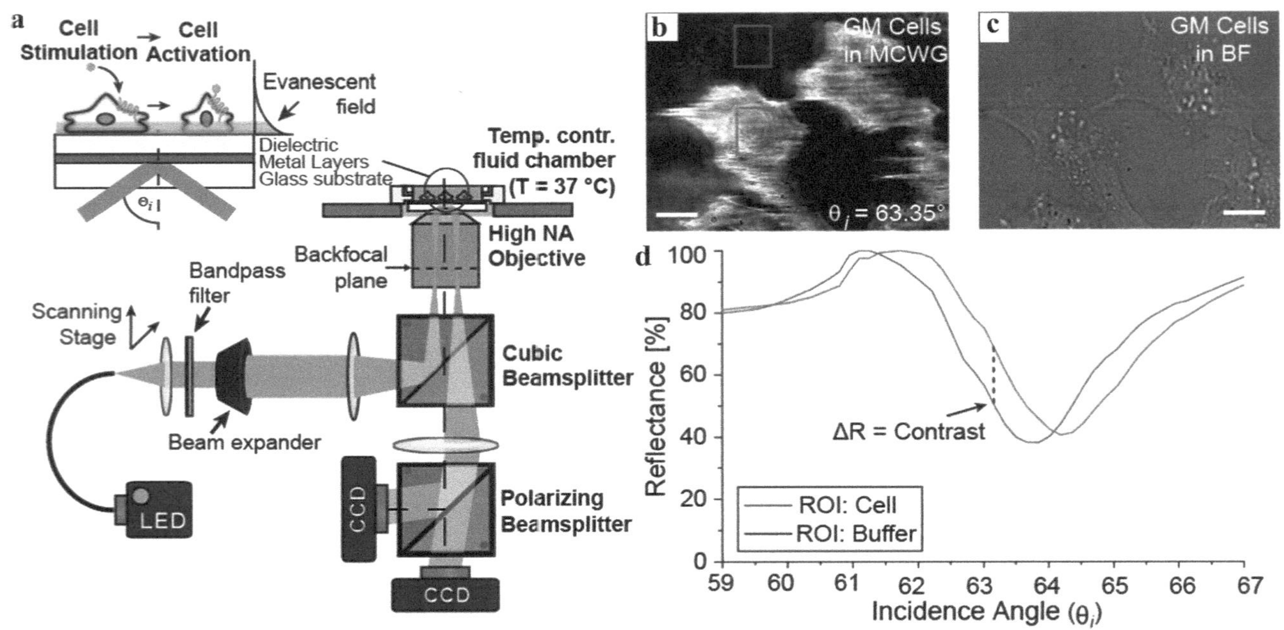

Fig. 9 (A) Schematic illustration of the PWR setup based on a high numerical aperture microscope objective. (B) Reflectance image of two glioblastoma cells (U251) cultured on top of the PWR sensor chip (scale bar = 20 μm, LED center wavelength = 470 nm incidence angle of 63.35 degrees). (C) A bright-field image of the same two cells. (D) Mean reflectance as a function of incidence angle for the two rectangular region of interests shown in (B), and the *vertical dotted line* indicates the operating angle (63.35 degrees) for fixed-angle measurements. *Reproduced with permission Sollradl, T.; Banville, F. A.; Frohlich, U.; Canva, M.; Charette, P. G.; Grandbois, M. Label-free visualization and quantification of single cell signaling activity using metal-clad waveguide (MCWG)-based microscopy. Biosens. Bioelectron.* ***2018,*** *100, 429–436. Copyright 2017, Elsevier B.V.*

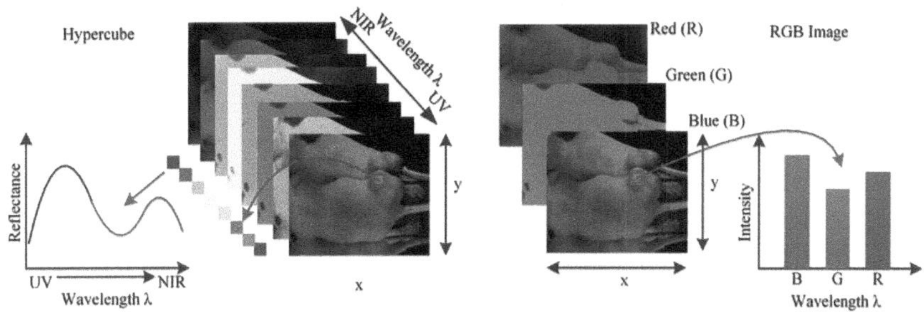

Fig. 10 Comparison between the hypercube and RGB images: hypercube is a three-dimensional dataset consisting of two-dimensional images at hundreds of wavelength bands and one-dimensional spectral signature of a pixel in the image (*lower left*). RGB image only records data in *red* (*light black in the print version*), *green* (*dark black in the print version*), and *blue* (*light white in the print version*) wavelengths bands and provides the intensity value of a pixel in the RGB image (*lower right*). *Reproduced with permission Lu, G.; Fei, B. Medical hyperspectral imaging: A review. J. Biomed. Opt.* ***2014,*** *19, 010901.*

transmission, scattering or fluorescence modes, which depends on the signal feature created by light-matter interaction. Multimodal HSI systems with high spatial resolution and high throughput characteristic spectra information, extend the application of HSI to micron and nanoscales, allowing biological analysis for a single cell. This section focuses on a brief overview of three HSI-based methods including dark-field hyperspectral imaging, hyperspectral Raman scattering and hyperspectral fluorescence imaging, and their recent achievements in the use of optical analysis of single cells.

3.1 Dark-field hyperspectral imaging

Dark-field hyperspectral imaging modality is achieved by integrating a hyperspectral imager with a standard optical microscope. The microscope works in the dark-field mode using oblique illumination, and the microscope objective only collects scattered light from the specimen, thereby creating high image contrast. The hyperspectral imager can acquire sufficient information in a measurement, including the spatial features and location of the sample, the spectrum of a single pixel and monochromatic images of single-bands, generating a three-dimensional data cube for further analysis. Fortunately, the spectral analysis tools and algorithms originally developed in the field of earth remote sensing can be easily converted to use in hyperspectral microscope imaging data processing, although there is scale difference between earth remote sensing and spectral microscopy. Early in 2006, Harris[52] has applied six spectral mapping methods derived from earth remote sensing to spectral microscopy data, namely, minimum distance classifier (MDC), spectral angle mapper (SAM), linear spectral unmixing (LSU), spectral feature fitting (SFF), matched filter (MF), and the mixture tuned matched filter (MTMF). Verebes et al.[53] built up a spectral library related to deformed red blood cells (RBCs) and white blood cells (WBCs) in an activated state, and utilized the SAM algorithm to evaluate the spectral similarity between sample spectra from the hyperspectral dark-field image and end member spectra from the spectral library. By mapping the spectral end members in the hyperspectral image and quantifying the relative abundance of each end member, three deformed shapes of RBCs (stacked erythrocytes, echinocytes, and irregularly shaped erythrocytes) were identified and neutrophils in an activated state were separated. Therefore, chemical variances and physical abnormalities in blood cells caused by pathological changes can be detected and characterized by dark-field microscopy based hyperspectral analysis. Conti et al.[54] for the first time developed a hyperspectral dark-field microscopy spectral library of human healthy RBCs. The spectral library consisted of eight spectral end members and their percent distribution in the optical image of erythrocytes was obtained by SAM analysis. As shown in Fig. 11, the end members of RBCs spectra were well matched with the spectral characteristics of the five main cell membrane components (phospholipid, cholesterol, hemoglobin, protoporphyrin, and spectrin). These detailed shape and spectral information about blood cells is helpful to identify cell damage caused by imbalance of molecular components and to evaluate the health and disease status of blood cells, as well as significant for assisting doctors in diagnosing diseases.

Noble metal nanoparticles (NPs) often act as optical probes in single-cell imaging due to their unique characteristic of local surface plasmon resonance (LSPR). The coupling of electromagnetic wave to LSPR will result in intense absorption and scattering of incident light at the resonant wavelength, which can be collected by dark-filed microscopy to investigate intracellular transport of these NPs. However, in traditional dark-field imaging, single NP and NPs clusters are difficult to distinguish only based on the independent spatial features. Hyperspectral dark-field microscopy provides more information about the spectral characteristics, so that NPs clusters can be identified by the spectral shift of the plasmon peak. Patskovsky et al.[55] observed the specific attachment of functionalized gold plasmonic nanoparticles on cancer stem cells through a supercontinuum white light laser-based hyperspectral dark-field imaging system, confirming that the monoclonal antibodies against human CD44 was in strong expression on MDA-MB-231 cells but no expression on BT-474 cells. Not as good as

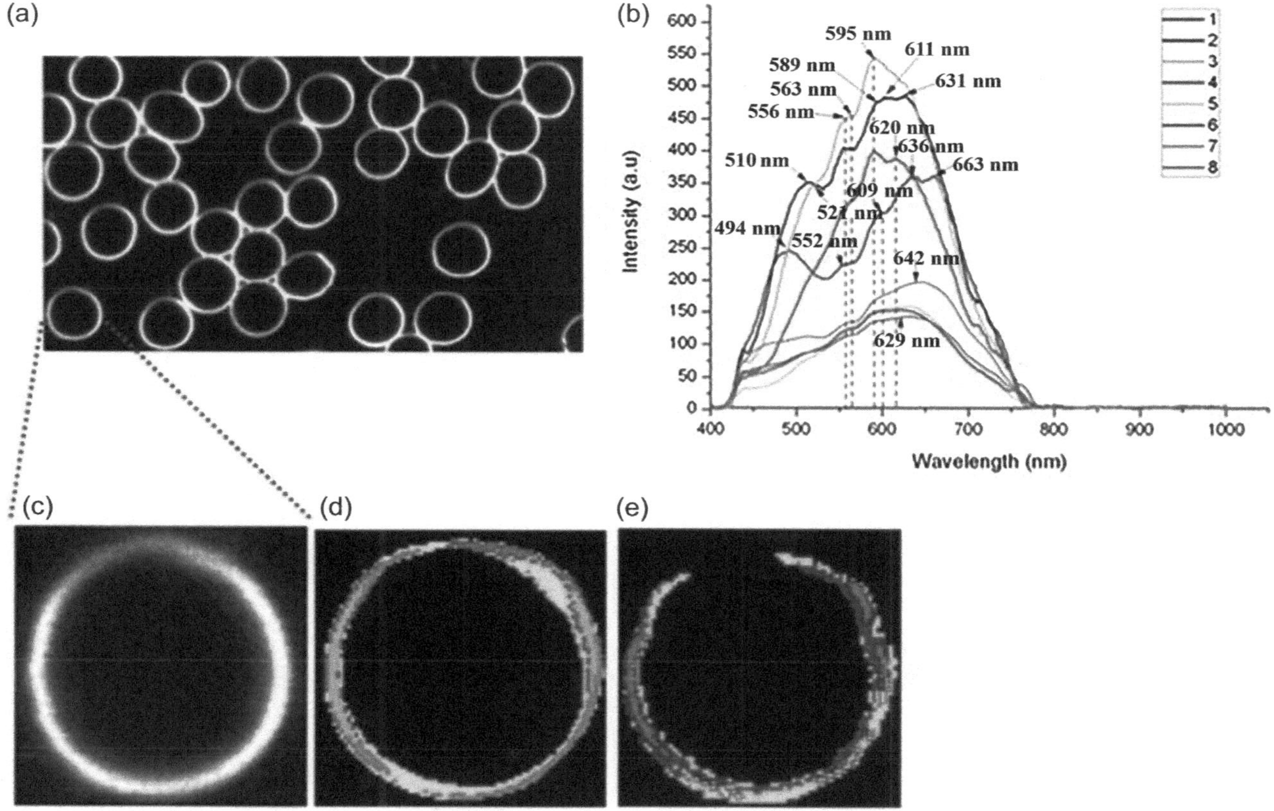

Fig. 11 (A) Hyperspectral dark-field image of erythrocyte. (B) Spectral library composed by eight spectral end members with arbitrary color code. (C) Enlarged image of a single red blood cell. (D) Mapping the spectral library in the optical image. (E) Mapping the five major cell membrane components (phospholipid, cholesterol, hemoglobin, protoporphyrin, and spectrin) in the optical image. *Adapted with permission Conti, M.; Scanferlato, R.; Louka, M.; Sansone, A.; Marzetti, C.; Ferreri, C., et al. Building up spectral libraries for mapping erythrocytes by hyperspectral dark field microscopy. Biomed. Spectrosc. Imaging* ***2016,*** *5, 175–184. Copyright 2016, IOS Press and the authors.*

the combination of a hyperspectral imager with a standard microscope, the filter-based detector system proposed by Patskovsky required more time-consuming scanning process and complex data analysis.

Microtubules (MTs) are the key protein filaments of the cytoskeleton, their polymerization and depolymerization are closely related to cell morphology maintenance, cell division, signal transduction, and intracellular transport.[56] While as gold nanoparticles (AuNPs) enter the cell, the size, and concentration of AuNPs may affect the polymerization of tubulin and the concomitant effects may induce undesirable protein conformational changes.[57] The AuNPs-induced protein conformational change resulted from the interaction of NPs with the tubulin–MTs was first observed by Choudhury et al.[57] using a hyperspectral dark-field microscopy along with flow cytometric and confocal laser scanning microscopic analysis. Mortimer et al.[58] detected and semi-quantified different metal nanoparticles, metal oxide nanoparticles and quantum dots in unicellular eukaryote *Tetrahymena thermophila* utilizing the dark-field enhanced hyperspectral imaging system, as shown in Fig. 12. Different types of NPs exhibit uniqueness in the spectrum, promoting the semi-quantitative study on cellular uptake of NPs by mapping spectral library in the HSI image. Wang et al.[59] quantified the cytosine modification on DNA at a single-cell level using the plasmonic nanoprobes silver nanoparticles (AgNPs) and AuNPs based on hyperspectral dark-field imaging, revealing the spatial distribution and quantitation of 5-carboxy cytosine at different stages of the cell cycle (shown in Fig. 13). Different from the process of hyperspectral data reported by Mortimer et al., Wang et al. used intensity and wavelength filters methods for removing the signal outliers before spectral mapping, which improved the quantification accuracy of hyperspectral dark-field imaging system and provided better quantification information about the epigenetic marks on single cells.

In addition to acting as the plasmonic nanoprobes functionalized for intracellular single-particles detection, noble metal NPs are also used as contrast-generating agents because of their large scattering cross-section to improve the imaging contrast and signal-to-noise ratio (SNR).[60–63] An optimized contrast-enhanced hyperspectral dark-field microscopy system for single cells imaging was constructed by Sen et al.,[63] which adopted the optical coherence tomography and used gold nanorods as exogenous contrast agents. Cui et al.[64] proposed a new single-cell optical clearing strategy to enhance the SNR of hyperspectral dark-field microscopy imaging without losing the primary cell content of nucleic acid and protein. On the basis of this method, as shown in Fig.14, transmembrane receptor protein

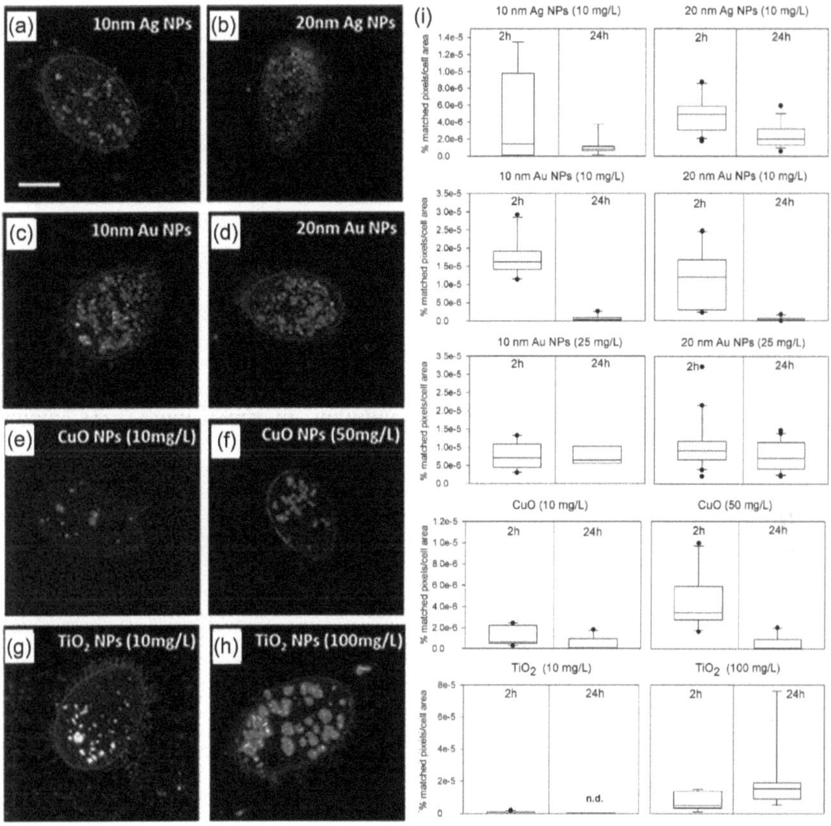

Fig. 12 Hyperspectral images and semi-quantification results of *Tetrahymena thermophila* uptake of NPs. (A–D) Cell incubated with 10 mg/L 10 nm AgNPs, 20 nm AgNPs, 10 nm AuNPs, and 20 nm AuNPs, respectively, for 2 h. (E–H) Cell incubated with 10 and 50 mg/L CuO NPs, 10 and 100 mg/L TiO_2 NPs, respectively, for 2 h; *red dots* (*gray in the print version*) on the images represent the intracellular location of NPs, and the scale bar = 10 μm; (I) *Box plots* showing semi-quantification results of internalized NPs after cell incubation for 2 and 24 h, expressed as a percentage of matching pixels per cell area. *Horizontal lines* are mean values, and *black dots* are outliers; Whiskers signify standard deviations ($n = 10$). *Reproduced with permission Mortimer, M.; Gogos, A.; Bartolome, N.; Kahru, A.; Bucheli, T. D.; Slaveykova, V. I. Potential of hyperspectral imaging microscopy for semi-quantitative analysis of nanoparticle uptake by protozoa. Environ. Sci. Technol.* ***2014****, 48, 8760–8767. Copyright 2014, American Chemical Society.*

HER2 and oncogenic mRNAs (ER and PR) in single cancer cells were quantitatively monitored and tracked with unprecedented spatiotemporal resolution.[64] Compared with NPs, oxygen nanobubbles (ONBs) exhibit higher SNR in the single-cell imaging, due to the light scattering caused by the large refractive index difference of core-shell structure.[65] As the

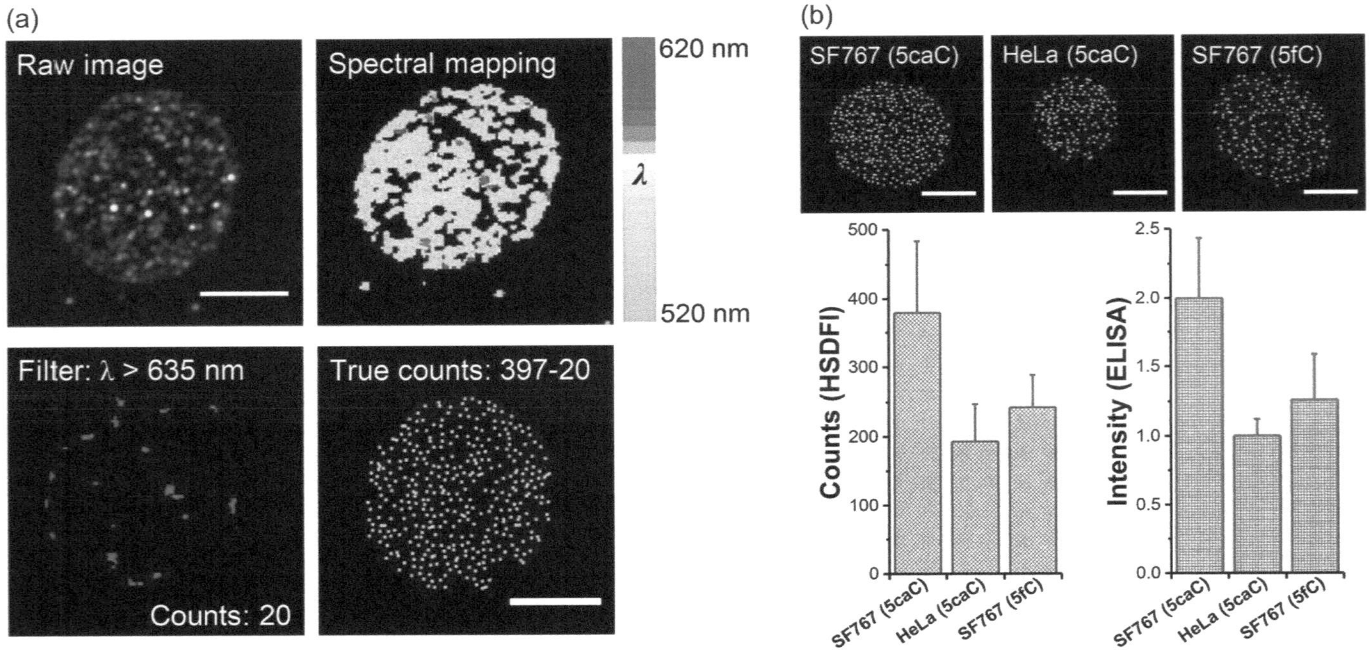

Fig. 13 Precise quantification of cytosine modifications with a filter function enabled by hyperspectral dark-field imaging. (A) The process of applying a filter function for AuNPs. (B) The *upper row* shows the binary images of the centroids of AuNPs targeting 5-carboxycytosine in SF767 cells and HeLa cells, and 5-formyl cytosine in SF767 cells, respectively. The *lower row* summarizes the counting results from 10 HSI image, which is validated by ELISA quantification in parallel ($n = 3$ replicates). Data are presented as mean with standard deviation and the scale bar = 10 μm. *Reproduced with permission Wang, X.; Cui, Y.; Irudayaraj, J. Single-cell quantification of cytosine modifications by hyperspectral dark-field imaging. ACS Nano* ***2015,*** *9, 11924–11932. Copyright 2015, American Chemical Society.*

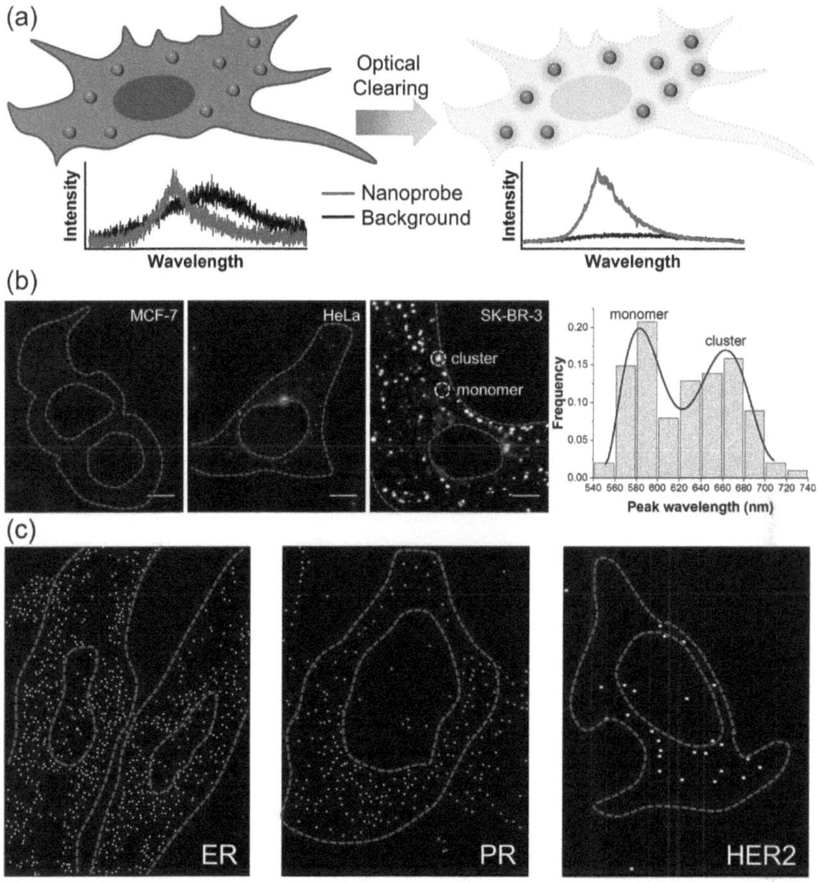

Fig. 14 (A) The process of single-cell optical clearing. (B) Dark-field image of HER2 in three cell lines and histogram of scattering peaks extracted from 15 HER2-positive SK-BR-3 cells (scale bars: 5 μm). (C) Three key mRNAs relating to gynecological cancers individually imaged in ER-expressing HeLa cells with improved SNR and centroid images reconstructed from raw images by hyperspectral processing (scale bars: 10 μm). *Adapted with permission Cui, Y.; Wang, X.; Ren, W.; Liu, J.; Irudayaraj, J. Optical clearing delivers ultrasensitive hyperspectral dark-field imaging for single-cell evaluation. ACS Nano* ***2016,*** *10, 3132–3143. Copyright 2016, American Chemical Society.*

simulation results of far-field scattering shown in Fig.15A–C, both the light field range and scattering cross section of ONBs are larger than those of AuNPs, ensuring the precise quantification and tracking of ONBs in single living cells (shown in Fig.15D–I). In summary, hyperspectral dark-field microscopy with the ability of single-pixel spectral analysis, breaks through the limitation that traditional imaging techniques only perform overall or

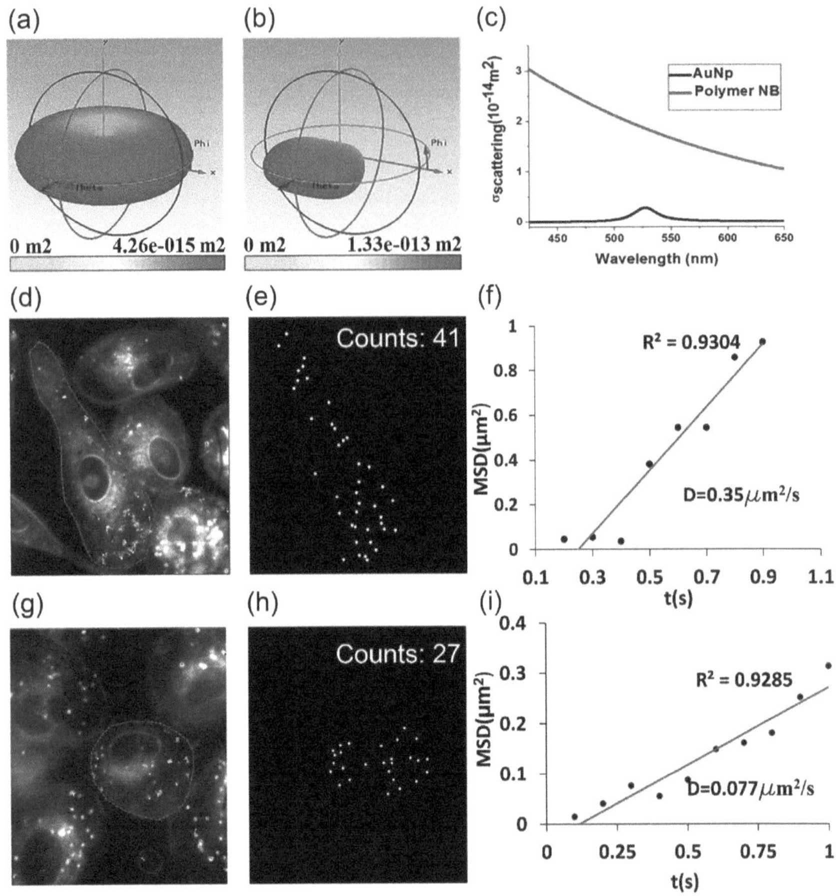

Fig. 15 Angular scattering pattern of (A) 40 nm AuNP and (B) 400 nm ONB in PBS. (C) Far-field scattering cross section of AuNP ONB with different wavelengths. (D) Tacking of 400 nm ONBs in PC3 human prostate cancer cell lines. (E) Quantification results of ONBs in cell after 2 h of incubation with 400 nm ONBs. (F) Diffusion coefficient of single 400 nm ONB based on fitting of the mean square displacement. (G) Tracking of 800 nm ONBs in PC3 human prostate cancer cell lines. (H) Quantification results of ONBs in cell after 2 h of incubation with 800 nm ONBs. (I) Diffusion coefficient of single 800 nm ONB based on fitting of the mean square displacement. *Adapted with permission Bhandari, P.; Wang, X.; Irudayaraj, J. Oxygen nanobubble tracking by light scattering in single cells and tissues. ACS Nano* ***2017,*** *11, 2682–2688. Copyright 2017, American Chemical Society.*

average measurements on cells or small samples. It is able to distinguish and quantify minute differences between different cell components, as well as map metal nanoparticles for cellular uptake without invasive sample preparation, enriching the toolbox for research of characterizing cellular components changes in complex biological environment, disease diagnosis and treatment, elucidating the biological effects' mechanisms of NPs and quantitative epigenetic screening at the single-cell level.

3.2 Hyperspectral Raman scattering

Raman spectroscopy, as a chemical fingerprint of molecules or materials, has attracted the most research attention in molecular detection and cell imaging. As early as more than a decade ago, a hyperspectral confocal SERS imaging system based on an AOTF was reported to identify the intracellular AgNPs labeled with cresyl violet.[66] AOTF-based Raman imaging records the entire SERS signals at each pixel in the two-dimensional image rather than the conventional Raman spectroscopy only offering point measurement. Heiner et al.[67] used the hyperspectral analysis to localize the intracellular reporter molecules carried by surface-enhanced hyper Raman scattering (SEHRS) nanoprobes in an endosomal system of macrophage cell. Compared with surface-enhanced Raman scattering, SEHRS, spontaneous two-photon excited Raman scattering, is more sensitive to the small environmental changes, and the spontaneous Raman scattering has higher spatial resolution owing to the shorter excitation wavelengths, beneficial for vibrational imaging of the complex biosystems. Stimulated Raman scattering (SRS), as well as with high sensitivity improved by using a high-frequency phase-sensitive detection scheme, combined with the hyperspectral imaging has been successfully applied for cellular chemical image and visual analysis of the intracellular distribution of small drug-like molecules.[68–70] The hyperspectral SRS imaging system for cellular imaging consists of two synchronized femtosecond lasers, an electro-optical modulator, a laser scanning microscope, a filter, a detector and a lock-in amplifier, referring to Fig. 16. Owing to the nonlinearity of stimulated Raman gain and loss in the overall excitation intensity, SRS spectroscopy shows great depth resolution for biological analysis of tissues and living cells,[71, 72] fostering the development of hyperspectral imaging combined with Raman spectroscopy for studying the multiple biomolecules and nanoparticles in three-dimensional cell cultures.[73] Huang et al. first observed the evident uptake competition and subcellular distribution of two different NPs (α-Fe_2O_3 and TiO_2) in the single cell by hyperspectral SRS imaging (shown in Fig. 17).[74] All the earlier-mentioned achievements achieved by hyperspectral SRS imaging, provide new avenues for in situ research of studying pharmacokinetics and understanding cellular dynamics with high spatial resolution and imaging speed.

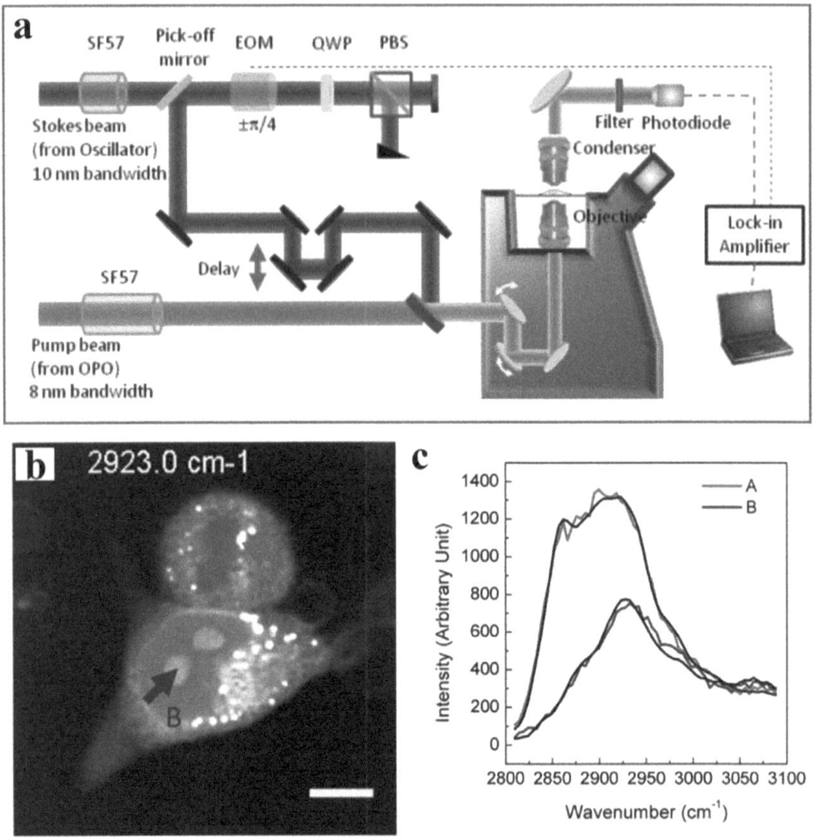

Fig. 16 (A) Schematic diagram of hyperspectral SRS imaging setup; SF57 is the glass rod for pulse stretching; *EOM*, electro-optical modulator; *QWP*, quarter wave plate; *PBS*, polarizing beam splitter. Two lasers with pulses stretched to ~2 ps are combined on a dichroic mirror and send into a laser scanning microscope for hyperspectral SRS imaging. (B) Hyperspectral imaging frame at 2923 cm^{-1} of two fixed HeLa cells. (C) SRS spectra at two individual pixels A, B in (B) and their spectral fitting (*black curve*) using five SRS basis spectra of saturated lipids DPPC, unsaturated lipids DOPC, cholesterol, protein and water; scale bar = 10 μm. *Part A: Reproduced with permission Fu, D.; Zhou, J.; Zhu, W. S.; Manley, P. W.; Wang, Y. K.; Hood, T. et al. Imaging the intracellular distribution of tyrosine kinase inhibitors in living cells with quantitative hyperspectral stimulated Raman scattering. Nat. Chem.* ***2014****, 6, 614–622. Copyright 2014, Nature Publishing Group. Part B and C: Adapted with permission Fu, D.; Holtom, G.; Freudiger, C.; Zhang, X.; Xie, X. S. Hyperspectral imaging with stimulated Raman scattering by chirped femtosecond lasers. J. Phys. Chem. B* ***2013****, 117, 4634–4640. Copyright 2012, American Chemical Society.*

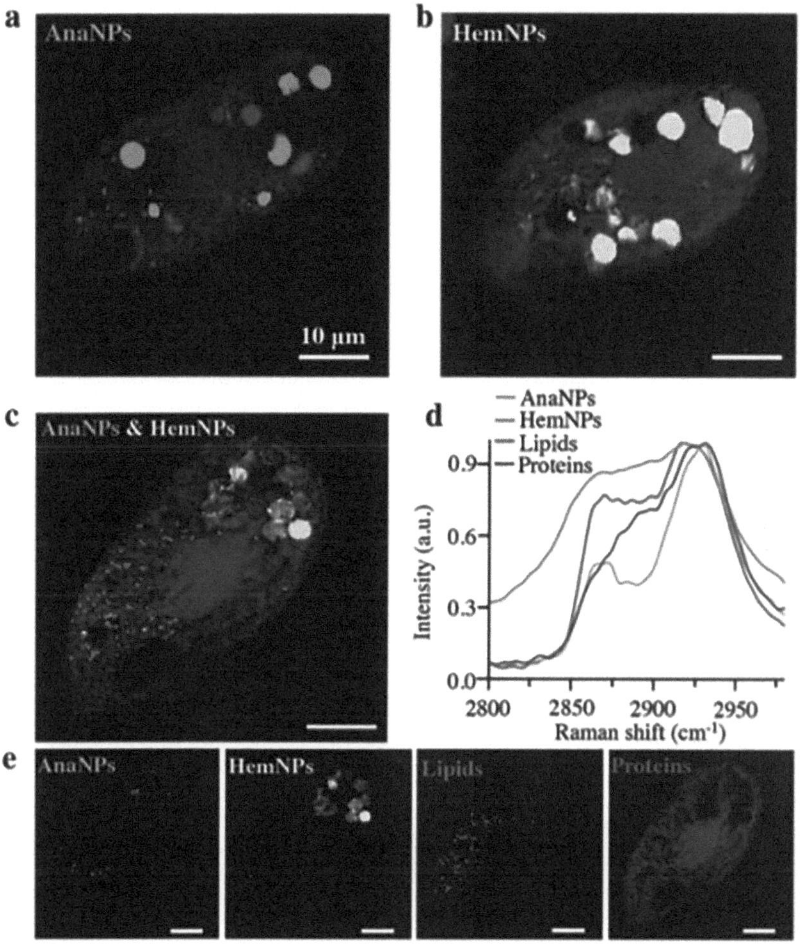

Fig. 17 Multivariate curve resolution (MCR)-retrieved concentration maps of in protozoan *Tetrahymena thermophila* pre-exposed to (A) TiO_2 NPs (13.2 mg Ti L^{-1}), (B) α-Fe_2O_3 NPs (30 mg Fe L^{-1}), and (C) a mixture of both. (D) MCR-retrieved Raman spectra and (E) MCR-reconstructed concentration images of AnaNPs (*green, dark gray in the print version*), HemNPs (*yellow, white in the print version*), lipids (*red, light gray in the print version*), and proteins (*purple, dark gray in the print version*) in (C). *Reproduced with permission Huang, B.; Yan, S.; Xiao, L.; Ji, R.; Yang, L.; Miao, A. J., et al. Label-free imaging of nanoparticle uptake competition in single cells by hyperspectral stimulated Raman scattering. Small* ***2018,*** *14, 1703246. Copyright 2017, WILEY-VCH Verlag GmbH & Co. KGaA, Weinheim.*

3.3 Hyperspectral fluorescence microscopy

A hyperspectral fluorescence microscope consists of a light source, an objective lens, a filter, and a hyperspectral camera.[75] Hyperspectral fluorescence data including microscopic images and fluorescence spectra can be used to analyze the biomolecular composition in individual cells. Hyperspectral confocal fluorescence microscope along with the robust MCR algorithm resolving the spectral overlap of fluorescence emission spectra has proven to be a powerful tool to localize and visualize distinctly photosynthesis-related pigments in cyanobacterial cells.[76, 77] The fluorescent probes such as multicolored fluorescent proteins, nanocrystals and organic fluorophores, play an important role in the study of intracellular protein localization and dynamics, but their applications in single cells suffer from the problem of emission spectra overlap. Elliott et al. built an optimized snapshot hyperspectral imaging system suitable for biological fluorescence microscopy, which was successfully used for real-time imaging of pancreatic β-cell dynamics in Hela cells with high spatial and spectral resolution.[78] Using the image mapping spectrometry (IMS),[79, 80] the image of triple-labeled cells acquired from the above imaging system clearly revealed the subcellular localization of ECFP, EGFP, and SYFP fluorescent proteins in the mitochondria, plasma membrane, and nucleus, respectively, while the color camera image only showed overlapping green shades. Recently, to break through the spectral resolution limitation of the microscope in the imaging of scattering samples in in-vivo measurements, a high spectral resolution hyperspectral multiphoton microscope with three excitation sources for multiplexed fluorescence emission has been developed to image cells labeled with multiple different fluorescent protein fusion constructs.[81]

Although the hyperspectral imaging methodology based on fluorescent microscopy has the inevitable disadvantage of photobleaching and phototoxicity, the great accomplishments achieved by this technology in single-cells research are worthy of recognition and encouragement. Challenges can be addressed by designing and developing advanced concepts and devices based on label-free technique. For instance, to overcome the drawback of flowing cytometry relying on the fluorescent labeling, Hiramarsu et al.[82] proposed a rapid-scan Fourier transform coherent anti-Stokes Raman scattering (FT-CARS) for directly probing intracellular molecules in a heterogeneous population of single cells. This FT-CARS flowing cytometry was successfully used to investigate photosynthetic dynamics and cellular heterogeneity of as taxanthin productivity in *Haematococcus lacustris* cells. Moreover, the relevant study of intracellular pigment distribution and content in

H. lacustris cells has been reported by Nogami et al.[83] The pigment content of every pixel in the hyperspectral microscopic image of *H. lacustris* was estimated by image-processing software combined with multivariate statistical analysis. Therefore, by virtue of the superiority of spectral imaging with high resolution and molecular fingerprinting, it is possible to develop a high-throughput, specific, visible, label-free hyperspectral Raman flowing cytometry for multimodal detection of cellular and intracellular changes. The above-mentioned SCA studies with use of hyperspectral dark-field microscopy mainly focus on stained or labeled cells. A simple and noninvasive characterization method suitable for rapid and direct analysis of unmarked single cells is greatly needed because we hope to achieve the original state of cells as much as possible with minimal intervention. To our knowledge, there are few reports on analyzing the components of unstained cells by using hyperspectral imaging. In our laboratory, a hyperspectral microscopy system was recently constructed by integrating a hyperspectral imager with a standard microscope, which allows for measurement in both bright-field and dark-field modes. Taking the human erythroleukemic cell line K562 as the object of investigation, we utilized the above-mentioned system to achieve the feature spectra of cellular membranes, cytoplasm and nucleus of the cell without any staining agent . The spectral imaging results obtained in bright-field mode are shown in Fig. 18. The spectral shape and intensity of different cellular composition are distinguishable. A straightforward comparison between the bright-field and dark-field images of the same cells indicates that the bright-field hyperspectral results are more informative than the dark-field results. More detailed information and more accurate characterization will be carried out in future work using well-designed segmentation model and algorithms.

4 Conclusion

Evanescent-wave optical biosensors with either SPR or WGR or PWR structure have been successfully applied to many single-cell analyses, including cell/substrate interactions, cell signaling, and cellular responses. SPR and WGR biosensors are well used to measure DMR of cells and different cellular responses. Compared with SPR biosensors, PWR biosensors are more suitable to monitor relatively thick objects because they have a deeper penetration depth than that of SPR biosensors, consequently perform higher sensitivity. There are two types of SPR imaging methods, based on a prism and a high numerical aperture, which can achieve high throughput and high

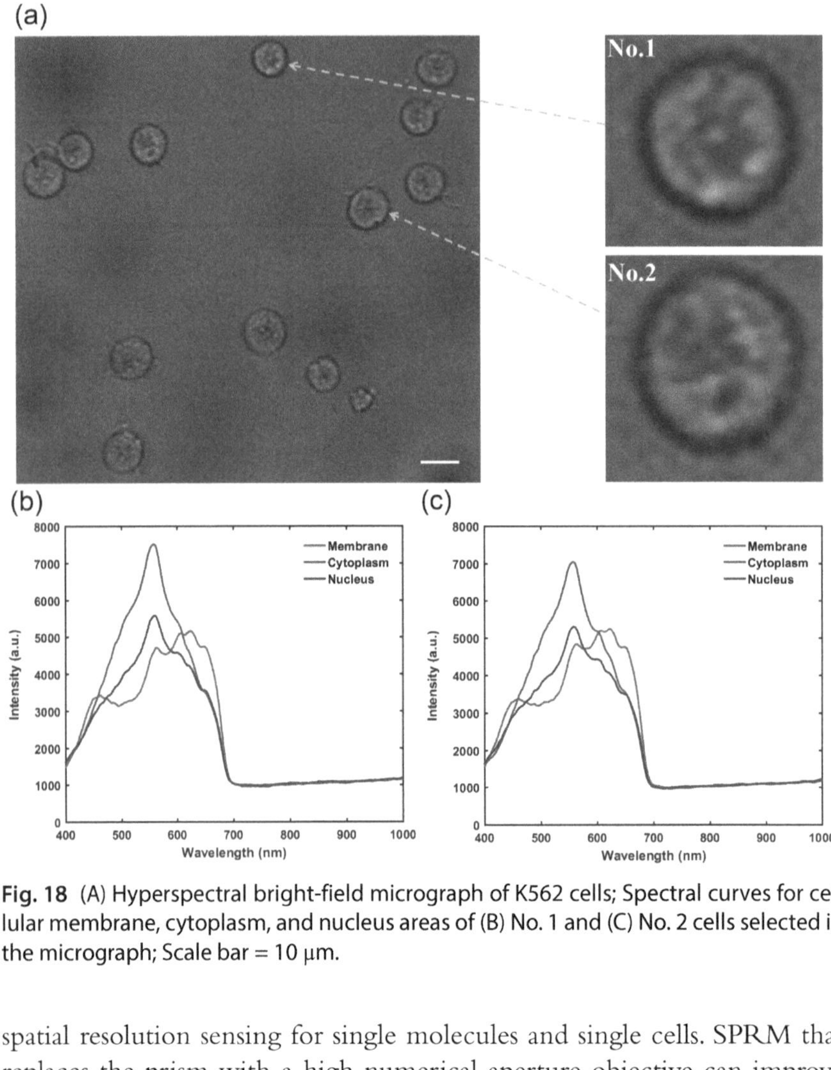

Fig. 18 (A) Hyperspectral bright-field micrograph of K562 cells; Spectral curves for cellular membrane, cytoplasm, and nucleus areas of (B) No. 1 and (C) No. 2 cells selected in the micrograph; Scale bar = 10 μm.

spatial resolution sensing for single molecules and single cells. SPRM that replaces the prism with a high numerical aperture objective can improve the spatial resolution to the diffraction limit but suffers from the limited dynamic range. What's more, there is a direct trade-off between spatial resolution and sample probing depth. Increasing the probe depth will decrease the resolution, and vice versa. In the future, these challenges are expected to be addressed through the development of well-designed spectral imaging methods using advanced nanofunctional materials and technologies.

Hyperspectral microscopic imaging techniques with dark-field microscopy, Raman scattering spectroscopy and fluorescence microscopy have been proven to be versatile tools for living cell sensing. They provide further insights for single-cell study, including classification of blood cells, understanding the chemical composition inside the cell, quantifying nanoparticles of cellular uptake, visual distribution of intracellular biomolecules and small drug-like molecules, and disease detection and treatment related to the individual cells. Hyperspectral dark-field microscopy has advantages of providing inherent imaging contrast, which is attributed to the scattering resulting from the uneven refractive index and thickness of cells and the large scattering cross section of nanoparticles, thereby widely used to analyze the unstained cells and characterize plasmonic nanoprobes in single cells. Hyperspectral simulated Raman scattering microscopy as an emerging label-free optical technique, is powerful for the chemical imaging of single cells because of its molecular fingerprint function. Hyperspectral fluorescent microscope can effectively resolve the overlapping spectra from multiple fluorescent labels, providing unprecedented specificity for identifying structures and cell types and promising for the research of complex in vivo studies. A common drawback of these three methods is that hyperspectral microscopy systems are more expensive than traditional spectrometers and CCD cameras due to their higher spatial and spectral resolution. And the extensive hypercube data acquired from those systems require precise statistical analysis and data processing, which is time-consuming. Besides, compared with Raman scattering and fluorescence imaging, the hyperspectral dark-field imaging approach does not suffer from autofluorescence, photobleaching, and phototoxicity, but is constrained by the diffraction limit. In the future, significant efforts are needed to develop advanced optical devices for increasing the speed of spectral imaging, and establish suitable theoretical models and optimal algorithms to improve the accuracy and speed of data processing. Furthermore, from the view of good compatibility of hyperspectral imaging and evanescent-wave sensing, it is expected to develop novel SPR-HSI or PWR-HSI integrated instruments with high tempo-spatial resolution, single-pixel spectral characterization and high-throughput screening capabilities. Such a powerful analytical tool could promote the development of single-cell analysis technology.

Acknowledgment

This work was supported by the National Natural Science Foundation of China (61931018, 61871365) and the Scientific Instrument Developing Project of the Chinese Academy of Sciences (YJKYYQ20180004).

References

1. Zhou, X. L.; Yang, Y.; Wang, S.; Liu, X. W. Surface plasmon resonance microscopy: from single-molecule sensing to single-cell imaging. *Angew. Chem. Int. Ed. Engl.* **2019**, *58*, 2–12.
2. Sun, D.; Cao, F.; Tian, Y.; Li, A.; Xu, W.; Chen, Q.; et al. Label-free detection of multiplexed metabolites at single-cell level via a SERS-microfluidic droplet platform. *Anal. Chem.* **2019**, *91*, 15484–15490.
3. Wang, J.; Geng, Y.; Shen, Y.; Shi, W.; Xu, W.; Xu, S. SERS-active fiber tip for intracellular and extracellular pH sensing in living single cells. *Sens. Act. B Chem.* **2019**, *290*, 527–534.
4. Damborsky, P.; Svitel, J.; Katrlik, J. Optical biosensors. *Essays Biochem.* **2016**, *60*, 91–100.
5. Chen, C.; Wang, J. Optical biosensors: An exhaustive and comprehensive review. *Analyst* **2020**, *145*, 1605–1628.
6. Fang, Y. Label-free biosensors for cell biology. *Int. J. Electrochem.* **2011**, *2011*, 460850.
7. Morris, J. D.; Payne, C. K. Microscopy and cell biology: New methods and new questions. *Annu. Rev. Phys. Chem.* **2019**, *70*, 199–218.
8. Mehta, N.; Shaik, S.; Devireddy, R.; Gartia, M. R. Single-cell analysis using hyperspectral imaging modalities. *J. Biomech. Eng.* **2018**, *140*, 020802.
9. Wood, R. W. On a remarkable case of uneven distribution of light in a diffraction grating spectrum. *Philos. Mag.* **1902**, *4*, 396–402.
10. Liedberg, B.; Nylander, C.; Lundstrom, I. Surface-plasmon resonance for gas-detection and biosensing. *Sens. Act.* **1983**, *4*, 299–304.
11. Yeatman, E.; Ash, E. A. Surface plasmon microscopy. *Electron. Lett.* **1987**, *23*, 1091–1092.
12. Huang, B.; Yu, F.; Zare, R. N. Surface plasmon resonance imaging using a high numerical aperture microscope objective. *Anal. Chem.* **2007**, *79*, 2979–2983.
13. Su, Y. W.; Wang, W. Surface plasmon resonance sensing: From purified biomolecules to intact cells. *Anal. Bioanal. Chem.* **2018**, *410*, 3943–3951.
14. Wang, W.; Yang, Y. Z.; Wang, S. P.; Nagaraj, V. J.; Liu, Q.; Wu, J.; et al. Label-free measuring and mapping of binding kinetics of membrane proteins in single living cells. *Nat. Chem.* **2012**, *4*, 846–853.
15. Giebel, K. F.; Bechinger, C.; Herminghaus, S.; Riedel, M.; Leiderer, P.; Weiland, U.; et al. Imaging of cell/substrate contacts of living cells with surface plasmon resonance microscopy. *Biophys. J.* **1999**, *76*, 509–516.
16. Wang, W.; Wang, S.; Liu, Q.; Wu, J.; Tao, N. Mapping single-cell-substrate interactions by surface plasmon resonance microscopy. *Langmuir* **2012**, *28*, 13373–13379.
17. Toma, K.; Kano, H.; Offenhaeusser, A. Label-free measurement of cell-electrode cleft gap distance with high spatial resolution surface plasmon microscopy. *ACS Nano* **2014**, *8*, 12612–12619.
18. Kreysing, E.; Hassani, H.; Hampe, N.; Offenhaeusser, A. Nanometer-resolved mapping of cell–substrate distances of contracting cardiomyocytes using surface plasmon resonance microscopy. *ACS Nano* **2018**, *12*, 8934–8942.
19. Yang, Y.; Yu, H.; Shan, X.; Wang, W.; Liu, X.; Wang, S.; et al. Label-free tracking of single organelle transportation in cells with nanometer precision using a plasmonic imaging technique. *Small* **2015**, *11*, 2878–2884.
20. Lu, J.; Li, J. Label-free imaging of dynamic and transient calcium signaling in single cells. *Angew. Chem. Int. Ed. Engl.* **2015**, *54*, 13576–13580.
21. Liu, X. W.; Yang, Y.; Wang, W.; Wang, S.; Gao, M.; Wu, J.; et al. Plasmonic-based electrochemical impedance imaging of electrical activities in single cells. *Angew. Chem. Int. Ed. Engl.* **2017**, *56*, 8855–8859.
22. Sereda, A.; Moreau, J.; Canva, M.; Maillart, E. High performance multi-spectral interrogation for surface plasmon resonance imaging sensors. *Biosens. Bioelectron.* **2014**, *54*, 175–180.

23. Zeng, Y.; Hu, R.; Wang, L.; Gu, D.; He, J.; Wu, S.-Y.; et al. Recent advances in surface plasmon resonance imaging: Detection speed, sensitivity, and portability. *Nanophotonics* **2017**, *6*, 1017–1030.
24. Knobloch, H.; von Szada Borryszkowski, G.; Woigk, S.; Helms, A.; Brehmer, L. Dispersive surface plasmon microscopy for the characterization of ultrathin organic films. *Appl. Phys. Lett.* **1996**, *69*, 2336–2337.
25. Liu, L.; He, Y.; Zhang, Y.; Ma, S.; Ma, H.; Guo, J. Parallel scan spectral surface plasmon resonance imaging. *Appl. Opt.* **2008**, *47*, 5616–5621.
26. Zeng, Y.; Wang, L.; Wu, S.Y.; He, J.; Qu, J.; Li, X.; et al. High-throughput imaging surface plasmon resonance biosensing based on an adaptive spectral-dip tracking scheme. *Opt. Exp.* **2016**, *24*, 28303–28311.
27. Zeng, Y. J.; Zhou, J.; Wang, X. L.; Cai, Z. W.; Shao, Y. H. Wavelength-scanning surface plasmon resonance microscopy:a novel tool for real time sensing of cell-substrate interactions. *Biosens. Bioelectron.* **2019**, *145*, 111717.
28. Chen, K.; Zeng, Y.; Wang, L.; Gu, D.; He, J.; Wu, S.Y.; et al. Fast spectral surface plasmon resonance imaging sensor for real-time high-throughput detection of biomolecular interactions. *J. Biomed. Opt.* **2016**, *21*, 127003.
29. Zeng, Y.; Wang, L.; Wu, S.Y.; He, J.; Qu, J.; Li, X.; et al. Wavelength-scanning SPR imaging sensors based on an acousto-optic tunable filter and a white light laser. *Sensors (Basel)* **2017**, 17.
30. Nielsen, T.; Schultz, R. A.; Wyatt, R.; Garner, H. R. Hyperspectral imaging: a novel approach for microscopic analysis. *Cytometry* **2001**, *43*, 239–247.
31. Novikova, I. V.; Smallwood, C. R.; Gong, Y.; Hu, D.; Hendricks, L.; Evans, J. E.; et al. Multimodal hyperspectral optical microscopy. *Chem. Phys.* **2017**, *498-499*, 25–32.
32. Ramsden, J. J.; Li, S. Y.; Heinzle, E.; Prenosil, J. E. Optical method for measurement of number and shape of attached cells in real time. *Cytometry* **1995**, *19*, 97–102.
33. Horvath, R.; Pedersen, H. C.; Skivesen, N.; Selmeczi, D.; Larsen, N. B. Monitoring of living cell attachment and spreading using reverse symmetry waveguide sensing. *Appl. Phys. Lett.* **2005**, *86*, 071101.
34. Horvath, R.; Cottier, K.; Pedersen, H. C.; Ramsden, J. J. Multidepth screening of living cells using optical waveguides. *Biosens. Bioelectron.* **2008**, *24*, 805–810.
35. Li, B.; Ju, H. Label-free optical biosensors based on a planar optical waveguide. *BioChip J.* **2013**, 7, 295–318.
36. Horváth, R.; Vörös, J.; Graf, R.; Fricsovszky, G.; Textor, M.; Lindvold, L. R.; et al. Effect of patterns and inhomogeneities on the surface of waveguides used for optical waveguide lightmode spectroscopy applications. *Appl. Phys. B* **2001**, 72, 441–447.
37. Fang, Y.; Ferrie, A. M.; Fontaine, N. H.; Mauro, J.; Balakrishnan, J. Resonant waveguide grating biosensor for living cell sensing. *Biophys. J.* **2006**, *91*, 1925–1940.
38. Cottier, K.; Horvath, R. Imageless microscopy of surface patterns using optical waveguides. *Appl. Phys. B* **2008**, *91*, 319–327.
39. Ferrie, A. M.; Wu, Q.; Fang, Y. Microfluidic resonant waveguide grating biosensor system for whole cell sensing. *Appl. Phys. Lett.* **2011**, *98*, 163703.
40. Fang, Y. Probing cancer signaling with resonant waveguide grating biosensors. *Exp. Opin. Drug Disc.* **2010**, *5*, 1237–1248.
41. Fang, Y.; Ferrie, A. M.; Herold, K. E.; Rasooly, A., Eds. *Resonant waveguide grating biosensor for cancer signaling*; CRC Press-Taylor & Francis Group: Florida, 2012; pp. 141–154.
42. Zaytseva, N.; Lynn, J. G.; Wu, Q.; Mudaliar, D. J.; Sun, H.; Kuang, P. Q.; et al. Resonant waveguide grating biosensor-enabled label-free and fluorescence detection of cell adhesion. *Sens. Act. B Chem.* **2013**, *188*, 1064–1072.
43. Sztilkovics, M.; Gerecsei, T.; Peter, B.; Saftics, A.; Kurunczi, S.; Szekacs, I.; et al. Single-cell adhesion force kinetics of cell populations from combined label-free optical biosensor and robotic fluidic force microscopy. *Sci. Rep.* **2020**, *10*, 61.

44. Salamon, Z.; Macleod, H. A.; Tollin, G. Coupled plasmon-waveguide resonators: A new spectroscopic tool for probing proteolipid film structure and properties. *Biophys. J.* **1997**, *73*, 2791–2797.
45. Pedersen, H. Optimization of metal-clad waveguide sensors. *Sens. Act. B Chem.* **2005**, *106*, 668–676.
46. Zhang, Z.; Lu, D. F.; Liu, Q.; Qi, Z. M.; Yang, L.; Liu, J. Wavelength-interrogated surface plasmon resonance sensor with mesoporous-silica-film-enhanced sensitivity to small molecules. *Analyst* **2012**, *137*, 4822–4828.
47. Wan, X. M.; Lu, D. F.; Gao, R.; Cheng, J.; Qi, Z. M. Metal-clad waveguide resonance sensor using a mesoporous TiO_2 thin film as the chemical sensitive core layer. *J. Phys. Chem. C* **2017**, *121*, 19173–19181.
48. Hotta, K.; Yamaguchi, A.; Teramae, N. Nanoporous waveguide sensor with optimized nanoarchitectures for highly sensitive label-free biosensing. *ACS Nano* **2012**, *6*, 1541–1547.
49. Sollradl, T.; Banville, F. A.; Chabot, V.; Canva, M.; Grandbois, M.; Charette, P. G. Metal clad waveguide (MCWG) based imaging using a high numerical aperture microscope objective. *Opt. Exp.* **2017**, *25*, 1666–1679.
50. Sollradl, T.; Banville, F. A.; Frohlich, U.; Canva, M.; Charette, P. G.; Grandbois, M. Label-free visualization and quantification of single cell signaling activity using metal-clad waveguide (MCWG)-based microscopy. *Biosens. Bioelectron.* **2018**, *100*, 429–436.
51. Sollradl, T.; Chabot, K.; Frohlich, U.; Canva, M.; Charette, P. G.; Grandbois, M. Monitoring individual cell-signaling activity using combined metal-clad waveguide and surface-enhanced fluorescence imaging. *Analyst* **2018**, *143*, 5559–5567.
52. Harris, A. T. Spectral mapping tools from the earth sciences applied to spectral microscopy data. *Cytom. Part. A* **2006**, *69*, 872–879.
53. Verebes, G. S.; Melchiorre, M.; Garcia-Leis, A.; Ferreri, C.; Marzetti, C.; Torreggiani, A. Hyperspectral enhanced dark field microscopy for imaging blood cells. *J. Biophoton.* **2013**, *6*, 960–967.
54. Conti, M.; Scanferlato, R.; Louka, M.; Sansone, A.; Marzetti, C.; Ferreri, C.; et al. Building up spectral libraries for mapping erythrocytes by hyperspectral dark field microscopy. *Biomed. Spectrosc. Imaging* **2016**, *5*, 175–184.
55. Patskovsky, S.; Bergeron, E.; Meunier, M. Hyperspectral darkfield microscopy of PEGylated gold nanoparticles targeting CD44-expressing cancer cells. *J. Biophoton.* **2015**, *8*, 162–167.
56. Zhang, Y. M.; Liu, J. H.; Yu, Q.; Wen, X.; Liu, Y. Targeted polypeptide-microtubule aggregation with cucurbit[8]uril for enhanced cell apoptosis. *Angew. Chem. Int. Ed. Engl.* **2019**, *58*, 10553–10557.
57. Choudhury, D.; Xavier, P. L.; Chaudhari, K.; John, R.; Dasgupta, A. K.; Pradeep, T.; et al. Unprecedented inhibition of tubulin polymerization directed by gold nanoparticles inducing cell cycle arrest and apoptosis. *Nanoscale* **2013**, *5*, 4476–4489.
58. Mortimer, M.; Gogos, A.; Bartolome, N.; Kahru, A.; Bucheli, T. D.; Slaveykova, V. I. Potential of hyperspectral imaging microscopy for semi-quantitative analysis of nanoparticle uptake by protozoa. *Environ. Sci. Technol.* **2014**, *48*, 8760–8767.
59. Wang, X.; Cui, Y.; Irudayaraj, J. Single-cell quantification of cytosine modifications by hyperspectral dark-field imaging. *ACS Nano* **2015**, *9*, 11924–11932.
60. Cognet, L.; Tardin, C.; Boyer, D.; Choquet, D.; Tamarat, P.; Lounis, B. Single metallic nanoparticle imaging for protein detection in cells. *Proc. Natl. Acad. Sci. U. S. A.* **2003**, *100*, 11350–11355.
61. Jun, Y. W.; Sheikholeslami, S.; Hostetter, D. R.; Tajon, C.; Craik, C. S.; Alivisatos, A. P. Continuous imaging of plasmon rulers in live cells reveals early-stage caspase-3 activation at the single-molecule level. *Proc. Natl. Acad. Sci. U. S. A.* **2009**, *106*, 17735–17740.

62. Lee, K.; Cui, Y.; Lee, L. P.; Irudayaraj, J. Quantitative imaging of single mRNA splice variants in living cells. *Nat. Nanotechnol.* **2014**, *9*, 474–480.
63. Sen, D.; SoRelle, E. D.; Liba, O.; Dalal, R.; Paulus, Y. M.; Kim, T. W.; et al. High-resolution contrast-enhanced optical coherence tomography in mice retinae. *J. Biomed. Opt.* **2016**, *21*, 066002.
64. Cui, Y.; Wang, X.; Ren, W.; Liu, J.; Irudayaraj, J. Optical clearing delivers ultrasensitive hyperspectral dark-field imaging for single-cell evaluation. *ACS Nano* **2016**, *10*, 3132–3143.
65. Bhandari, P.; Wang, X.; Irudayaraj, J. Oxygen nanobubble tracking by light scattering in single cells and tissues. *ACS Nano* **2017**, *11*, 2682–2688.
66. Wabuyele, M. B.; Yan, F.; Griffin, G. D.; Vo-Dinh, T. Hyperspectral surface-enhanced Raman imaging of labeled silver nanoparticles in single cells. *Rev. Sci. Instrum.* **2005**, *76*, 063710.
67. Heiner, Z.; Guhlke, M.; Zivanovic, V.; Madzharova, F.; Kneipp, J. Surface-enhanced hyper Raman hyperspectral imaging and probing in animal cells. *Nanoscale* **2017**, *9*, 8024–8032.
68. Fu, D.; Zhou, J.; Zhu, W. S.; Manley, P. W.; Wang, Y. K.; Hood, T.; et al. Imaging the intracellular distribution of tyrosine kinase inhibitors in living cells with quantitative hyperspectral stimulated Raman scattering. *Nat. Chem.* **2014**, *6*, 614–622.
69. Fu, D.; Lu, F. K.; Zhang, X.; Freudiger, C.; Pernik, D. R.; Holtom, G.; et al. Quantitative chemical imaging with multiplex stimulated Raman scattering microscopy. *J. Am. Chem. Soc.* **2012**, *134*, 3623–3626.
70. Hill, A. H.; Fu, D. Cellular imaging using stimulated Raman scattering microscopy. *Anal. Chem.* **2019**, *91*, 9333–9342.
71. Freudiger, C. W.; Min, W.; Saar, B. G.; Lu, S.; Holtom, G. R.; He, C.; et al. Label-free biomedical imaging with high sensitivity by stimulated Raman scattering microscopy. *Science* **2008**, *322*, 1857–1861.
72. Fu, D.; Holtom, G.; Freudiger, C.; Zhang, X.; Xie, X. S. Hyperspectral imaging with stimulated Raman scattering by chirped femtosecond lasers. *J. Phys. Chem. B* **2013**, *117*, 4634–4640.
73. Kallepitis, C.; Bergholt, M. S.; Mazo, M. M.; Leonardo, V.; Skaalure, S. C.; Maynard, S. A.; et al. Quantitative volumetric Raman imaging of three dimensional cell cultures. *Nat. Commun.* **2017**, *8*, 14843.
74. Huang, B.; Yan, S.; Xiao, L.; Ji, R.; Yang, L.; Miao, A. J.; et al. Label-free imaging of nanoparticle uptake competition in single cells by hyperspectral stimulated Raman scattering. *Small* **2018**, *14*, 1703246.
75. Pu, H.; Lin, L.; Sun, D. W. Principles of hyperspectral microscope imaging techniques and their applications in food quality and safety detection: A review. *Compr. Rev. Food Sci. F.* **2019**, *18*, 853–866.
76. Sinclair, M. B.; Haaland, D. M.; Timlin, J. A.; Jones, H. D. T. Hyperspectral confocal microscope. *Appl. Opt.* **2006**, *45*, 6283–6291.
77. Vermaas, W. F. J.; Timlin, J. A.; Jones, H. D. T.; Sinclair, M. B.; Nieman, L. T.; Hamad, S. W.; et al. In vivo hyperspectral confocal fluorescence imaging to determine pigment localization and distribution in cyanobacterial cells. *Proc. Natl. Acad. Sci. U. S. A.* **2008**, *105*, 4050–4055.
78. Elliott, A. D.; Gao, L.; Ustione, A.; Bedard, N.; Kester, R.; Piston, D. W.; et al. Real-time hyperspectral fluorescence imaging of pancreatic beta-cell dynamics with the image mapping spectrometer. *J. Cell Sci.* **2012**, *125*, 4833–4840.
79. Gao, L. A.; Kester, R. T.; Hagen, N.; Tkaczyk, T. S. Snapshot image mapping spectrometer (IMS) with high sampling density for hyperspectral microscopy. *Opt. Exp.* **2010**, *18*, 14330–14344.

80. Cui, Q.; Park, J.; Theodore Smith, R.; Gao, L. Snapshot hyperspectral light field imaging using image mapping spectrometry. *Opt. Lett.* **2020**, *45*, 772–775.
81. Bares, A. J.; Mejooli, M. A.; Pender, M. A.; Leddon, S. A.; Tilley, S.; Lin, K.; et al. Hyperspectral multiphoton microscopy for in vivo visualization of multiple, spectrally overlapped fluorescent labels. *Optica* **2020**, 7, 1587–1601.
82. Hiramatsu, K.; Ideguchi, T.; Yonamine, Y.; Lee, S.; Luo, Y. Z.; Hashimoto, K.; et al. High-throughput label-free molecular fingerprinting flow cytometry. *Sci. Adv.* **2019**, *5*. eaau0241.
83. Nogami, S.; Ohnuki, S.; Ohya, Y. Hyperspectral imaging techniques for the characterization of Haematococcus pluvialis (Chlorophyceae). *J. Phycol.* **2014**, *50*, 939–947.

CHAPTER 7

Future directions of biosensors for single-cell analysis from the perspective of instrument developments

Huiwen Tan[a,b], Xuantao Su[c,d], Deyong Chen[a,b,e], Junbo Wang[a,b,e,*], Yao Lu[f,*], and Jian Chen[a,*]
[a]State Key Laboratory of Transducer Technology, Aerospace Information Research Institute, Chinese Academy of Sciences, Beijing, China
[b]School of Future Technology, University of Chinese Academy of Sciences, Beijing, China
[c]School of Microelectronics, Shandong University, Jinan, Shandong, China
[d]Advanced Medical Research Institute, Shandong University, Jinan, Shandong, China
[e]School of Electronic, Electrical and Communication Engineering, University of Chinese Academy of Sciences, Beijing, China
[f]Department of Biotechnology, Dalian Institute of Chemical Physics, Chinese Academy of Sciences, Dalian, Liaoning, China

1 Introduction

Within the past several decades, tons of biosensors have been developed to estimate single-cell biophysical and/or biochemical properties. However, the majority of these measurement approaches is still confined within laboratories and cannot be transferred to commercially available instruments. Without the key step of instrumentation, the impacts of these biosensors are severely compromised. To address this issue, historical developments of hematology analyzer as a representative of clinical instruments and flow cytometry as a representative of research instruments are summarized and compared in the following sections. Then future directions of biosensors for single-cell analysis are suggested from the perspective of instrumentation.

2 Instrumentation of hematology analyzer (Table 1)

The first hematology analyzer was patented in 1953 by Wallace H. Coulter and commercialized by The Coulter Electronics as *Model A*. In this prototype, a cell suspension was drawn through an aperture positioned between

*Co-Corresponding authors.

Biosensors for Single-Cell Analysis
https://doi.org/10.1016/B978-0-323-89841-6.00002-5

Table 1 Historical development of hematology analyzer.

Manufacturer	Instrument	Methodology	Parameter
Coulter electronics	Model A	DC resistance	
Coulter electronics	Model S Plus	DC resistance	Three-part differential of WBC
Coulter electronics	STKs	DC/AC imcpedance + light scattering	Five-part differential of WBC
TOA medical electronics	Sysmex NE-8000	DC/AC impedance + selective lysis	Five-part differential of WBC
Abbott laboratories	Cell-DYN 3000	Multiangle light scattering	Five-part differential of WBC
Roche	Argos 5-Diff	DC resistance + selective staining and lysis + light absorbance	Five-part differential of WBC
Beckman Coulter	DxH series	DC/AC impedance + light scattering	Five-part differential of WBC
Sysmex	XN L Series	Light scattering + fluorescence detection	Five-part differential of WBC + immature granulocytes
Mindray	BC-7500 CRP	Light scattering + fluorescence detection	Five-part differential of WBC + immature granulocytes + high-fluorescence cells

two electrodes and corresponding changes in electrical resistance were measured as voltage pulses. The number of pulses corresponded to the number of cells, and the height of each pulse was proportional to the volume of the cell. In 1977, the Coulter Electronics introduced an automatic hematology analyzer (*Model S Plus*) to realize a three-part differential of white blood cells based on direct current resistances. In this instrument, large (> 160 fL), intermediate (90–160 fL), and small (< 90 fL) size groups represented neutrophils and eosinophils; monocytes and mononuclear cells; lymphocytes and basophils, respectively.

As to five-part differential with neutrophils, monocytes, eosinophils, basophils, and lymphocytes, *STKs* of the Coulter Electronics functioned as a 22-parameter automated hematology analyzer that could process 100 samples per hour. Three different technologies were used in this instrument: volumetric resistance; conductivity using high frequency electromagnetic energy to analyze cellular physical and chemical compositions; and a monochromatic laser beam to determine both cell size and structures. *Sysmex NE-8000* of the TOA Medical Electronics was a 23-parameter automated hematology analyzer (120 samples per hour) that used both radio frequency to obtain information on nuclear sizes and densities of cells and direct current for size estimation. Eosinophils and basophils were determined in separate detection blocks after using cell-specific lysing reagents, followed by direct current detection. Furthermore, *Cell-DYN 3000* of Abbott measured 22 parameters based on multiangle light scattering to realize five-part differential relying on differences of sizes, granularities, and internal complexities. *Cobas Argos 5-Diff* of Roche analyzed 26 parameters based on resistance, cytochemical staining of eosinophils, and optical absorbance. Basophils were determined in a separate channel by using a lyses followed by impedance counting.

Currently, hematology analyzers can quantify not only five-part differential of white blood cells but also additional parameters (e.g., nucleated red blood cells, large unstained cells, hematopoietic progenitor cells, immature granulocytes, granularity index, and lymph index), which can help to identify atypical populations of cells to aid in diagnosis of certain conditions of diseases. More specifically, *DxH series* of the Beckman Coulter enables a five-part differential of white blood cells based on DC/AC impedance and light scattering. *XN L series* of Sysmex enables a five-part differential of white blood cells with immature granulocytes quantified based on the combination of light scattering and fluorescence detection. *BC-7500 CRP* of Mindray realizes a five-part differential with immature granulocytes and high-fluorescence cells classified using light scattering and fluorescence detection.

3 Instrumentation of flow cytometer (Table 2)

The first instrumentation of the flow cytometer was developed in 1968 by German developer and manufacturer Partec under the name *impulsecytophotometer ICP11*. At that time, absorption methods were still widely favored over fluorescence methods. In 1971, Bio/Physics Systems commercialized

Table 2 Instrumentation of flow cytometer.

Manufacturer	Instrument	Key performance
Partec	Impulscytophotometer ICP 11	Measurement of optical absorbance relying on PMT
Bio/Physics Systems	Cytofluorograph	Measurement of forward scatter light, green and red fluorescence relying on PMT
Becton Dickinson	FACS II	Measurement of forward scatter light and green fluorescence with subsequent cell sorting
Coulter Electronics	EPICS C analyzer	Measurement of three fluorescence simultaneously relying on PMT
Becton Dickinson	FACScan	Measurement of three fluorescence simultaneously relying on PMT
Partec	PAS-III	Measurement of seven fluorescence simultaneously relying on PMT
Becton Dickinson	LSR II	Measurement of 14 fluorescence simultaneously relying on PMT
BD Biosciences	FACSymphony A5	Measurement of 50 parameters simultaneously relying on PMT
Fluidigm	Helios	Measurement of more than 100 parameters simultaneously relying on mass spectroscopy
Luminex	ImageStreamX Mark II	Image collection with 12 channels relying on CCD

the brand name of *Cytofluorograph*, which was equipped with a 633-nm He-Ne laser and a 488-nm argon ion laser, enabling the measurement of forward scatter light, green and red fluorescence emissions. In 1974, the Becton Dickinson commercialized *FACS II* featured with optical characterization and subsequent sorting by including the droplet deflection technology.

In 1980s, competitions started among flow cytometer manufacturers concerning the simultaneous analysis capacity of fluorescence. For instance, *EPICS C* from the Coulter Electronics and *FACScan* from the Becton-Dickinson allowed up to three color analysis by including an additional arc lamp or a 15 mW air-cooled argon laser source, respectively. In 1990s, the capacity of flow cytometers was increased to measure seven fluorescence simultaneously, which was demonstrated by *PAS-III* of the Partec. In 2000s, *LSR II* of the Becton-Dickinson was capable of detecting 14 fluorescence simultaneously. Currently, state-of-art fluorescent flow cytometers can simultaneously measure up to 50 parameters based on 9 lasers and a maximal 10 PMTs per laser (*FACSymphony A5* of BD Biosciences).

Besides the commercialization of fluorescent flow cytometers, a mass flow cytometer was first commercialized by the DVS Sciences as the band name of *CyTOF* where metal-tagged antibodies were used to replace fluorescent labels to significantly decrease minimal background noises from signal overlaps. The state-of-art mass flow cytometer is commercialized by Fluidigm with the name of *Helios*, which is capable of quantifying more than 100 parameters simultaneously. Meanwhile, an imaging flow cytometer was first commercialized by the Amnis with the brand name of *ImageStream100* by collecting digital images of traveling cells as a combination of microscopy and flow cytometry. The state-of-art imaging flow cytometer is commercialized by Luminex with the brand name of *ImageStreamX Mark II* which is capable of obtaining up to 12 channels of cellular imagery at a speed of 5000 cells/s.

4 Future directions

A retrospective study of hematology analyzers and flow cytometers indicates two different roadmaps of innovations for biosensors enabling single-cell analysis. For the developments of hematology analyzers, it follows the path of demand-driven innovation. To meet the demands of clinical hematology (e.g., effectively classification of white bloods), three generations of hematology analyzers were developed to realize three-part differential, five-part differential, and higher than five-part differential of white blood cells. To meet these clinical demands, technical innovations including DC/AC impedance, light scattering, fluorescence detection were witnessed within the developments of hematology analyzers. The successful commercialization of hematology analyzers indicate one future direction of biosensors, which is motivated by unmet biological demands at the single-cell level.

As to the developments of flow cytometers, it follows the path of the technology-driven innovation. To extend technical capabilities of flow cytometers, the numbers of simultaneously measured parameters have been increased from 1, to 3, to 7, to 14, and even 50. With the inclusion of mass and imaging flow cytometers, measurement capabilities of flow cytometers have been further extended. However, if the true clinical demands of flow cytometry (e.g., diagnosis and management of leukemia) are taken into consideration, measurements of nine fluorescence are still commonly used in hospitals. The successful commercialization of flow cytometry reveals the second future direction of biosensors, which is motivated by nonstop improvements of technical capabilities at the single-cell level.

Acknowledgment

The authors acknowledge financial supports from the National Natural Science Foundation of China (Grant No. 61825107, 61922079, 62001042), Key Project (QYZDB-SSW-JSC011), Instrument Development, Youth Innovation Promotion Association, and Interdisciplinary Innovation Team of Chinese Academy of Sciences.

Index

Note: Page numbers followed by *t* indicate tables.

R

S

Kolibri 360 GmbH
Ferdinand-Jühlke-Straße 7
99095 Erfurt, Deutschland
produktsicherheit@kolibri360.de